Timo Ströhle

Inverse dynamics of underactuated flexible mechanical systems governed by quasi-linear hyperbolic partial differential equations

Schriftenreihe des Instituts für Mechanik
Karlsruher Institut für Technologie (KIT)

Band 11

Herausgeber:
Prof. Dr.-Ing. habil. Peter Betsch
Prof. Dr.-Ing. habil. Thomas Seelig

Eine Übersicht aller bisher in dieser Schriftenreihe erschienenen Bände finden Sie am Ende des Buches.

Inverse dynamics of underactuated flexible mechanical systems governed by quasi-linear hyperbolic partial differential equations

by
Timo Ströhle

KIT Scientific Publishing

Karlsruher Institut für Technologie
Institut für Mechanik

Inverse dynamics of underactuated flexible mechanical systems governed by quasi-linear hyperbolic partial differential equations

Zur Erlangung des akademischen Grades eines Doktor-Ingenieurs von der KIT-Fakultät für Bauingenieur-, Geo- und Umweltwissenschaften des Karlsruher Instituts für Technologie (KIT) genehmigte Dissertation

von Timo Ströhle, M.Sc.

Tag der mündlichen Prüfung: 3. Juli 2023
Hauptreferent: Prof. Dr.-Ing. habil. Peter Betsch
Korreferentin: Prof. Dr.-Ing. habil. Sigrid Leyendecker

Impressum

KIT Scientific Publishing

Karlsruher Institut für Technologie (KIT)
KIT Scientific Publishing
Straße am Forum 2
D-76131 Karlsruhe

KIT Scientific Publishing is a registered trademark of Karlsruhe Institute of Technology.
Reprint using the book cover is not allowed.

www.ksp.kit.edu

Print on Demand 2024 – Gedruckt auf FSC-zertifiziertem Papier

ISSN 2363-4936
ISBN 978-3-7315-1336-0
DOI 10.5445/KSP/1000165695

Abstract

This work is about the inverse dynamics of underactuated flexible mechanical systems governed by quasi-linear hyperbolic partial differential equations subjected to time-varying *Dirichlet* boundary conditions that are enforced by unknown, spatially disjunct, hence non-collocated *Neumann* boundary conditions. The governing equations are first derived abstractly in more detail before various mechanical systems are presented aligning with the postulated formulation. Therefore, geometrically exact theories, governing the large motion of slender structures such as strings and beams but also general *three*-dimensional continua, will be derived.

Commonly, initial boundary value problems that occur in non-linear structural dynamics are solved by applying sequential space-time discretization methods. This approach is therefore typically based on a discretization in space by means of finite elements, followed by an appropriate discretization in time mostly based on finite differences. A brief survey of this type of sequential space-time integration methods for the initial boundary value problem at hand, is given first in context of the direct dynamics problem. I.e. the pure *Neumann* boundary problem will be considered first, before different possibilities of imposing *Dirichlet* boundary conditions in general, will be discussed afterwards. Based on this, the inverse dynamics problem will be introduced in context of spatially discrete mechanical systems subjected to rheonomic holonomic servo-constraints. A detailed analysis of this type of constrained systems aims to elaborate the fundamental distinctions of servo-constraints to classical contact-constraints. Consequences thereof, regarding the construction of numerically stable integration methods, will be addressed likewise before numerous selected examples will be given.

Due to the highly restrictive applicability of solving the inverse dynamics problem sequentially in space and time, an in-depth analysis of the underlying initial boundary value problem is being pursued. Especially by exposing the underlying hyperbolic structure of the governing partial differential equations, further insights into the problem at hand are anticipated. Enlighting the resulting mechanisms of wave propagation within continuous structures will pave the way to a numerically stable integration of the inverse dynamics problem. Thus, e.g. a method that is based on the integration of partial differential equations along characteristic manifolds is presented. This motivates the development of novel Galerkin methods that can be presented in this work as well.

These newly established methods will be applied to the feed-forward control problem of various mechanical systems. In addition to that, the novel simultaneous space-time integration methods are adopted to flexible multibody systems such as e.g. the cooperative control of a rigid body through several flexible strings or the control of the end-effector of a rotational arm consisting of rigid and flexible arms.

Selected numerical examples are given underlining the relevance of the proposed simultaneous space-time integration.

Kurzfassung

Diese Arbeit befasst sich mit der inversen Dynamik unteraktuierter, flexibler, mechanischer Systeme, welche durch quasi-lineare hyperbolische partielle Differentialgleichungen beschrieben werden können. Diese Gleichungnen, sind zeitlich veränderlichen *Dirichlet*-Randbedingungen unterworfen, welche durch unbekannte, räumlich disjunkte, also nicht kollokierte *Neumann*-Randbedingungen erzwungen werden. Die zugrundeliegenden Gleichungen werden zunächst abstrakt hergeleitet, bevor verschiedene mechanische Systeme vorgestellt werden können, die mit der eingangs postulierten Formulierung übereinstimmen. Hierzu werden geometrisch exakte Theorien hergeleitet, welche in der Lage sind große Bewegungen schlanker Strukturen wie Seile und Balken, aber auch ganz allgemein, dreidimensionaler Festkörper zu beschreiben.

In der Regel werden Anfangs-Randwertprobleme, die in der nichtlinearen Strukturdynamik auftreten, durch Anwendung einer sequentiellen Diskretisierung in Raum und Zeit gelöst. Diese Verfahren basieren für gewöhnlich auf einer räumlichen Diskretisierung mit finiten Elementen, gefolgt von einer geeigneten zeitlichen Diskretisierung, welche meist auf finiten Differenzen beruht. Ein kurzer Überblick über derartige sequentielle Integrationsverfahren für das vorliegende Anfangs-Randwertproblem wird zunächst anhand der direkten Formulierung des Problems gegeben werden. D.h. es wird zunächst das reine *Neumann*-Randproblem betrachtet, bevor anschließend ganz allgemein, verschiedene Möglichkeiten zur Einbindung etwaiger *Dirichlet*-Randbedingungen diskutiert werden. Darauf aufbauend wird das Problem der inversen Dynamik im Kontext räumlich diskreter mechanischer Systeme, welche rheonom-holonomen Servo-Bindungen unterliegen, eingeführt. Eine ausführliche Untersuchung dieser Art von gebundenen Systemen soll die grundlegenden Unterschiede zwischen Servo-Bindungen und klassischen Kontakt-Bindungen herausarbeiten. Die daraus resultierenden Folgen für die Entwicklung geeigneter numerisch stabiler Integrationsverfahren können dabei ebenfalls angesprochen werden, bevor zahlreich ausgewählte Beispiele vorgestellt werden können.

Aufgrund der sehr eingeschränkten Anwendbarkeit der sequentiellen Lösung der inversen Dynamik in Raum und Zeit, wird eine eingehende Analyse des vorliegenden Anfangs-Randwertproblems unternommen. Vor allem durch die Freilegung der hyperbolischen Struktur der zugrundeliegenden partiellen Differentialgleichungen werden sich weitere Einblicke in das vorliegende Problem erhofft. Die Erforschung der daraus resultierenden Mechanismen der Wellenausbreitung in kontinuierlichen Strukturen öffnet die Tür zur Entwicklung numerisch stabiler Integrationsverfahren für die inverse Dynamik. So kann unter anderem eine Methode vorgestellt werden, die auf der Integration der partiellen Differentialgleichungen entlang charakteristischer Mannigfaltigkeiten beruht. Dies regt zu der Entwicklung neuartiger Galerkinverfahren an, die ebenfalls in dieser Arbeit vorgestellt werden können.

Diese neu entwickelten Methoden können anschließend auf die Steuerung verschiedener mechanischer Systeme angewendet werden. Darüber hinaus können die neuartigen Integrationsverfahren auch auf flexible Mehrkörpersysteme übertragen werden. Angeführt seien hier beispielsweise die kooperative Steuerung eines an mehreren flexiblen Seilen aufgehängten starren Körpers oder die Steuerung des Endeffektors eines flexiblen mehrgliedrigen Schwenkarms.

Ausgewählte numerische Beispiele verdeutlichen die Relevanz der hier vorgeschlagenen, in Raum und Zeit simultanen Integration des vorliegenden Anfangs-Randwertproblems.

Acknowledgments

The work presented in this thesis has been carried out while working as a research associate and doctoral candidate at the Institute of Mechanics (IFM) at the Karlsruhe Institute of Technolgy (KIT) during the period 2017-2023. This research was funded by the Deutsche Forschungsgemeinschaft (DFG, German Research Foundation), which is gratefully acknowledged. Besides that, many people have supported me over the last six years on my way and I think this is the right place to say thank you.

First and foremost, I would like to thank my supervisor Prof. Peter Betsch for the confidence he has placed in me. He encouraged me to undertake this scientific venture and supported me whenever it was necessary. His way to think is unique and has always been inspiring to me. I am sincerely grateful for the open-minded atmosphere he has created at the Institute. I also have enjoyed our daily lunch-routine a lot.

I would also like to thank Prof. Sigrid Leyendecker (FAU Erlangen-Nürnberg) for her commitment acting as external examiner. Her interest in my work has been encouraging and has lead to valuable improvements.

Special thanks go to Jonas Hund for proofreading the manuscript and for being always a faithful companion. Moreover, I am very grateful to all of my colleagues at the IFM creating the best working atmosphere that one could ask for. I will always remember the time I have spent with all of you.

Finally and most importantly, I want to thank my family and friends for all their continuous support and unconditional love.

Karlsruhe, December 2023 Timo Ströhle

Contents

1. Introduction

How does a flexible body need to be excited on a prescribed boundary, such that the motion on a spatially disjunct boundary can partly be specified? This seemingly simple question will be pursued in this work and problems, that arise by attempting to solve such inverse dynamics problems of spatially continuous systems using classical numerical solution strategies, will be pointed out.

1.1. Motivation and formulation of the problem

In the sequel, we will focus on infinite-dimensional systems governed by second-order quasi-linear hyperbolic partial differential equations of the form

$$\boldsymbol{A} \cdot \partial_t^2 \boldsymbol{x} - \mathrm{div}_s(\boldsymbol{B} \cdot \partial_s \boldsymbol{x}) = \boldsymbol{C} \, . \tag{1.1}$$

Since we are interested in quasi-linear equations, we explicitly allow the coefficients $\boldsymbol{A}$, $\boldsymbol{B}$ and $\boldsymbol{C}$ to depend on the variables $s \in S \subset \mathbb{R}^d$ and $t \in \mathcal{T} \subset \mathbb{R}_+$, as well as on the solution $\boldsymbol{x} : \Omega \mapsto \mathbb{R}^d$ and their first partial derivatives itself, such that

$$\boldsymbol{A}, \boldsymbol{B} : \bar{\Omega} \mapsto \mathbb{R}^{d,d} \quad \text{and} \quad \boldsymbol{C} : \bar{\Omega} \mapsto \mathbb{R}^d \quad \text{where} \quad \bar{\Omega} = \Omega \cup \{(\boldsymbol{x}, \partial_s \boldsymbol{x}, \partial_t \boldsymbol{x}) : \Omega \mapsto \mathbb{R}^d\}.$$

For convenience, the space-time domain $\Omega = S \times \mathcal{T}$ has been introduced, where $S \subset \mathbb{R}^\alpha$ represents the spatial and $\mathcal{T} \subset \mathbb{R}_+$ the temporal domain. Note that the spatial dimension $d > 1$, does not necessarily coincide with the dimension of the spatial variable $\alpha > 1$.

In context of the classical direct formulation of the problem, the task is to find a solution $\boldsymbol{x} : \Omega \mapsto \mathbb{R}^d$ satisfying Eq. (1.1), that additionally is subjected to some given *Neumann* boundary conditions on $\partial\Omega_f = \partial S_f \times \mathcal{T}$

$$\boldsymbol{B} \partial_s \boldsymbol{x}|_{(s,t) \in \partial\Omega_f} = \boldsymbol{f}(s, t) \tag{1.2}$$

and (time-varying) *Dirichlet* boundary conditions on $\partial\Omega_\eta = \partial S_\eta \times \mathcal{T}$[1]

$$\boldsymbol{x}|_{(s,t) \in \partial\Omega_\eta} = \boldsymbol{\gamma}(s, t) \tag{1.3}$$

[1] In the special case of spatially *one*-dimensional systems, i.e. $d = 1$, we define $\partial S_f \equiv \{0\}$ and $\partial S_\eta \equiv \{L\}$, for some $L \in \mathbb{R}$.

as well as initial conditions on $\partial\Omega_0 = S \times \{0\}$

$$\boldsymbol{x}|_{(s,t)\in\partial\Omega_0} = \boldsymbol{x}_0 \qquad \text{and} \qquad \partial_t \boldsymbol{x}|_{(s,t)\in\partial\Omega_0} = \boldsymbol{v}_0 . \tag{1.4}$$

Note that for the direct problem the conditions

$$\boldsymbol{B}\partial_s \boldsymbol{x}|_{(s,t)\in\partial\Omega_\eta} = \boldsymbol{\eta}(s,t) \tag{1.5}$$

on $\partial\Omega_\eta$ need to be undetermined, i.e. *Dirichlet* and *Neumann* boundaries need to be disjunct.

In contrast to that, the inverse problem seeks to find a function $\boldsymbol{f} : \partial\Omega_f \mapsto \mathbb{R}^d$ on $\partial\Omega_f$ by additionally imposing *Neumann* boundary conditions on $\partial\Omega_\eta = \partial S_\eta \times \mathcal{T}$ as given by

$$\boldsymbol{B}\partial_s \boldsymbol{x}|_{(s,t)\in\partial\Omega_\eta} = \boldsymbol{\eta}(\partial_t^2 \boldsymbol{x}, \partial_t \boldsymbol{x}, \boldsymbol{x}, t) \ . \tag{1.6}$$

Note that the function $\boldsymbol{\eta}$ in general might be given in differential form.

Whereas in the direct problem given *Dirichlet* boundary conditions (1.3) on $\partial\Omega_\eta$ are enforced by the unknown function $\boldsymbol{\eta} : \partial\Omega_\eta \mapsto \mathbb{R}^d$, acting on the same boundary $\partial\Omega_\eta$, the inverse problem seeks to find the unknown function $\boldsymbol{f} : \partial\Omega_f \mapsto \mathbb{R}^d$ on $\partial\Omega_f$ enforcing (1.3). This unknown *Neumann*-boundary conditions can then be considered as Lagrange multipliers, enforcing the given time-variant *Dirichlet*-boundary conditions.

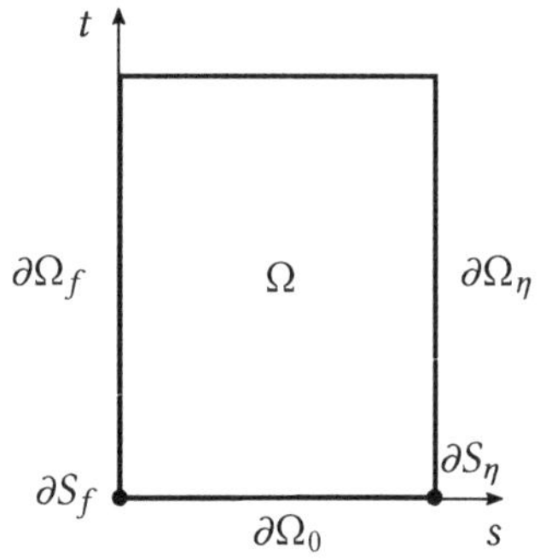

Figure 1.1.: Illustration of the space-time domain for $\alpha = 1$.

For the spatially *one-* and *two*-dimensional case, i.e. $\alpha = 1$ and $\alpha = 2$, the space-time domain takes the form of a plane and cylinder, respectively (cf. Fig. 1.1 and Fig. 1.2, for a visualization thereof, including all relevant boundaries).

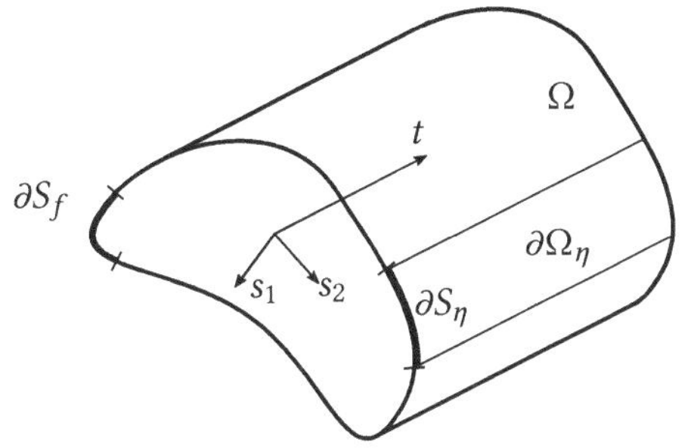

Figure 1.2.: Illustration of the space-time domain for $\alpha = 2$.

Commonly initial boundary value problems are solved by applying sequential space-time integration methods. For that purpose, the underlying partial differential equation is commonly transformed into a system of ordinary differential equations by applying e.g. the finite element method based on a spatial discretization (see, e.g. [34, 62]). Boundary conditions might be taken into account either by choosing suitable test function spaces or by imposing feasible geometric constraints to the configuration space. The motion of the system is then formulated in terms of redundant coordinates on a specified constraint manifold. The resulting spatially discrete system can be solved by applying appropriate time integration schemes mostly based on finite differences[2].

In principle the same approach can be applied to the inverse dynamics problem. However, since the constraints do not have collocation property in the sense that the corresponding constraint forces do not refer to the same location as the constraints does, they need to be distinguished from classical geometric constraints. From a geometrical point of view, the constraint forces are non-(ideal) orthogonal to the constraint manifold. This specific type of constraints are frequently referred to as servo-, control- or program-constraints (see [79], [64] and [22], respectively).

Due to this inherent non-standard construction of the constraint realization the resulting differential-algebraic system of equations is known to be characterized either by internal dynamics (see e.g. [56],[82] and [87]) or by excessively high differentiation index along with excessively high demands on the prescribed constraint function (see e.g. [28], [25], [13], [87], [5] and [63]). In the former case, the internal dynamics tend to be unstable depending on the spatial discretization. This hinders a numerical stable integration of the problem at hand. Therefore, it is inevitable to identify the unstable internal dynamics and carry out relevant analysis thereof (cf. e.g. [17], [85], [14]). In the latter case, appropriate

[2] Actually, instead of solving a hyperbolic problem, elliptic problems are solved successively in time

index reduction techniques need to be applied prior to the numerical time integration since a numerically stable integration of the resulting differential algebraic system of equations is depending significantly on the differentiation index (see e.g. [24] and [4]). Unfortunately it turns out, that both, the differentiation index as well as the demands on the smoothness of the imposed constraint increases with the refinement of the spatial discretization.

The highly restrictive applicability of solving the inverse dynamics problem sequentially in space-time motivates this work. Our goal is to develop novel solution strategies circumventing the issues outlined above. Therefore, analysing the governing equations in more detail seams to be inevitable. In particular, exposing the underlying hyperbolic structure will lead to a much better understanding of how information flow in physical time dependent problems. The causality of the solution, i.e. that the solution depends on the past but not on the future, will provide valuable insights into the problem and will turn out to be crucial for the inverse dynamics of infinite-dimensional systems. Consequently, simultaneous space-time discretization approaches may prove to be the natural choice for the solution of the inverse dynamics problem. This may confirm the famous quotation of *T.J.R. Hughes* and *G.M. Hulbert* as given by

> *'In fact, the more one thinks about it, the more apparent it becomes that the ubiquitous semidiscrete approach is conceptually confining, even schizophrenic.'* [52]

This will pave the way of successfully developing numerical integration methods that are capable to solve the inverse dynamics of infinite-dimensional systems.

Representative numerical examples including geometrically exact formualtions of slender structures undergoing large deformations, such as ropes and beams as well as general nonlinear continuum formulation demonstrate the capabilities of the newly devised space-time integration methods. Also flexible multibody systems will be addressed, underpinning the relevance of the proposed methods.

1.2. Outline

The rest of this work is outlined as follows.

Chapter 2 aims to introduce various mechanical systems aligning with the initial boundary value problem postulated abstractly in Section 1.1. In particular geometrically exact formulations governing the motion of slender structures such as strings and beams are derived in Section 2.1 and Section 2.2, respectively. Section 2.3 demonstrates that the general theory of *three*-dimensional continua aligns with the postulated formulation as well.

In **Chapter 3** classical solution strategies for the inverse dynamics problem that are based on a sequential space-time integration, are addressed. Therefore, Section 3.1 aims to give a brief survey of sequential space-time integration methods for the initial boundary value problem at hand. This will be done first in context of the direct dynamics problem. In particular, the pure *Neumann* boundary problem will be considered first, before different possibilities of imposing *Dirichlet* boundary conditions in general, are discussed afterwards. Based on this, the inverse dynamics problem is introduced in context of spatially finite-dimensional mechanical systems subjected to time-varying servo-constraints in Section 3.2. A detailed analysis of such constrained systems aims to elaborate the fundamental distinctions of servo-constraints to classical contact-constraints. Consequences thereof, regarding the construction of numerically stable integration methods, will be addressed likewise. Numerous selected examples will be given.

Due to the highly restrictive applicability of solving the inverse dynamics problem sequentially in time, the governing partial differential equation is analyzed in **Chapter 4** in more detail. In particular, exposing the underlying hyperbolic structure of the governing partial differential equations anticipates to gain more insights into the problem at hand. Enlighting resulting mechanisms of how information flows within continuous structures will pave the way to solve the inverse dynamics problem at hand numerically stable. Therefore, in Section 4.1 and in Section 4.2, novel numerical methods are presented, that are based on a simultaneous space-time discretization. Several numerical examples are presented in Section 4.3, underpinning the relevance of the proposed methods.

Chapter 5 aims to apply the newly established theory of simultaneous space-time integration to the control problem of flexible multibody systems. Therefore, a strategy to solve the cooperative control problem of a rigid body actuated through several flexible ropes is established in Section 5.1 before a rotational arm model consisting of *two* rigid and *one* flexible arm, is analyzed in Section 5.2. Accompanying numerical examples are underlining the relevance of the proposed methods in context of flexible mechanical systems.

Conclusions will be drawn in **Chapter 6**.

2. Infinite-dimensional mechanical systems

In Sec. 1.1, the inverse dynamics of underactuated flexible mechanical systems have been formulated in terms of quasi-linear hyperbolic partial differential equations subjected to time-varying *Dirichlet* boundary conditions that are enforced by unknown, spatially disjunct, i.e. non-collocated *Neumann* boundary conditions. The present chapter aims to derive various mechanical systems, aligning with this specific initial boundary value problem. Therefore, slender structures such as strings and beams as well as general *three-dimensional* continua are analyzed in Sec. 2.1, Sec. 2.2 and Sec. 2.3, respectively.

2.1. Geometrically exact string formulation[1]

This section aims to introduce a geometrically exact formulation for strings undergoing finite deformations, that aligns with the framework postulated in Sec. 1.1. An exhaustive account of the underlying geometrically exact description of strings can be found in the book by Antman [7]. To identify material points lying on the reference curve of the string, we make use of the arc-length in the reference configuration $s \in S = [0, L] \subset \mathbb{R}$ (cf. Fig. 2.1). In this connection, we assume a stress-free reference configuration in which the string has length L.

The placement of a material point $s \in S$ at any time $t \in \mathcal{T} = [0, \infty)$ is characterized by the deformation map $\boldsymbol{r}(s,t) : S \times \mathcal{T} \mapsto \mathbb{R}^d$, where $d \in \{1, 2, 3\}$ is the spatial dimension. In the sequel it proves convenient to introduce the space-time domain $\Omega = S \times \mathcal{T} \subset \mathbb{R}^2$.

Balance of linear momentum gives rise to the underlying equations of motion which take the form of the following system of partial-differential equations

$$\partial_s \boldsymbol{n}(s,t) + \boldsymbol{b}(s,t) = (\rho A)(s)\partial_t^2 \boldsymbol{r}(s,t) \tag{2.1}$$

[1] This section partly reproduces [93]

for all $(s,t) \in \Omega$. Here, $\boldsymbol{n}(s,t) : \Omega \mapsto \mathbb{R}^d$ represents the contact force at $(s,t) \in \Omega$, $\boldsymbol{b}(s,t) : \Omega \mapsto \mathbb{R}^d$ denotes the body force per unit reference length at $(s,t) \in \Omega$ and $(\rho A)(s) : S \mapsto \mathbb{R}_+$ is the mass density per unit reference length at s. In this connection, ρ stands for the density per unit reference volume and A refers to the cross-sectional area of the string in the reference configuration. The body force per unit reference length is given by

$$\boldsymbol{b}(s,t) = (\rho A)(s)\boldsymbol{g} \tag{2.2}$$

where $\boldsymbol{g}$ is the vector of gravitational acceleration.

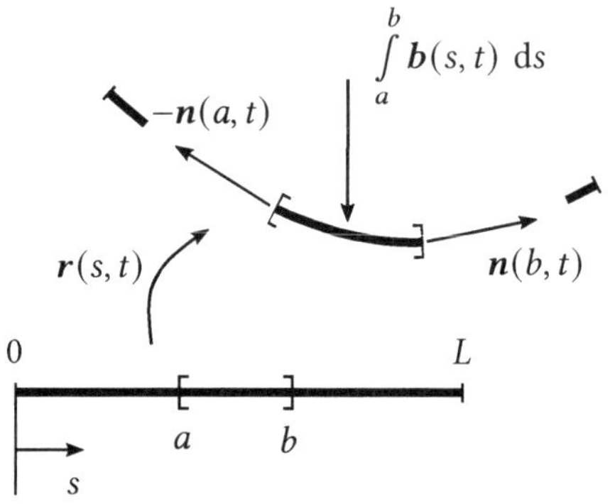

Figure 2.1.: Illustration of the geometrically exact string.

Since the string has no bending stiffness, the contact forces in the string need to be oriented tangentially to the string. That is,

$$\boldsymbol{n}(s,t) = N(s,t)\frac{\partial_s \boldsymbol{r}(s,t)}{\|\partial_s \boldsymbol{r}(s,t)\|} . \tag{2.3}$$

The tension $N(s,t)$ at $(s,t) \in \Omega$ is determined by a frame-indifferent constitutive relation of the form

$$N(s,t) = \hat{N}(\nu(s,t), s) \tag{2.4}$$

where the stretch at $(s,t) \in \Omega$ is defined by

$$\nu(s,t) = \|\partial_s \boldsymbol{r}(s,t)\| . \tag{2.5}$$

The total actual length of the string follows from

$$l(t) = \int_S \nu(s,t)\ \mathrm{d}s . \tag{2.6}$$

Note that the stretch represents the local ratio of the deformed to the reference length of the string. A physically meaningful elastic constitutive law needs to satisfy the restrictions $\hat{N}(\nu,s) \to \infty$ as $\nu \to \infty$ and $\hat{N}(\nu,s) \to -\infty$ as $\nu \to 0$. In addition to that, we assume a natural reference configuration implying $\hat{N}(1,s) = 0$, i.e. vanishing tension in the reference configuration.

Example 2.1 (Hyperelastic material model). *Subsequently, we apply a hyperelastic constitutive model for a homogeneous rope based on the stored energy* $\hat{U}(v) = \frac{EA}{4}(v^2 - 2\ln v - 1)$. *This stored energy represents a simple model for uniaxial rubber-type material behavior. The tension in the string follows from* $\hat{N}(v) = \hat{U}'(v)$ *leading to the constitutive relationship*

$$\hat{N}(v) = \frac{EA}{2}(v - v^{-1}) \ . \tag{2.7}$$

Linearization of the last equation about the natural reference configuration yields the linearized constitutive relationship

$$\hat{N}_{\text{lin}}(v) = EA(v - 1) \ . \tag{2.8}$$

Thus, EA can be regarded as axial stiffness in the reference configuration.

For the inverse dynamics problem, we assume that the motion of the elastic string is partially specified at its boundary at $s = L$ (Fig. 2.2). That is, the placement of the free end of the string at $s = L$ is specified for all times, leading to the boundary conditions

$$\boldsymbol{r}(L,t) = \boldsymbol{\gamma}(t) \quad \text{and} \quad \boldsymbol{n}(L,t) = \boldsymbol{0} \quad \forall\, t \in T \tag{2.9}$$

where $\boldsymbol{\gamma} : \mathcal{T} \mapsto \in \mathbb{R}^d$ is a prescribed function of time.

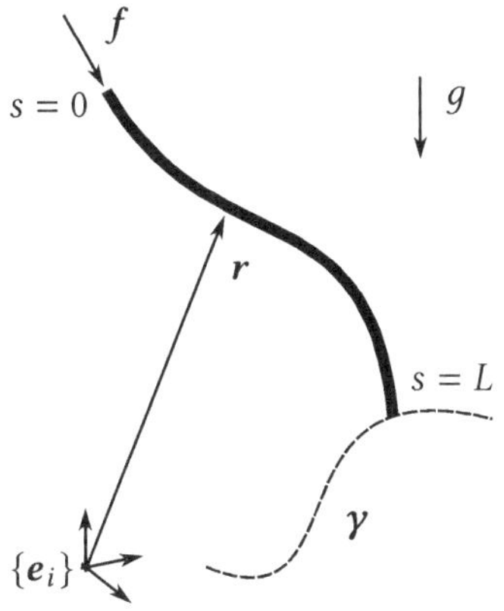

Figure 2.2.: Illustration of the inverse dynamics problem of the geometrically exact string.

The objective is to find the actuating force vector $\boldsymbol{f} : \mathcal{T} \mapsto \mathbb{R}^d$ acting on the boundary at $s = 0$ causing the partially prescribed motion of the string at $s = L$. The corresponding boundary condition can be written as

$$\boldsymbol{n}(0,t) = \boldsymbol{f}(t) \quad \forall\, t \in \mathcal{T} \ . \tag{2.10}$$

The resulting feedforward control problem gives rise to the initial boundary value problem postulated in Sec. 1.1. For the geometrically exact string formulation, the problem may be summarized as follows: Find $\boldsymbol{r}(s,t)$ and $\boldsymbol{f}(t)$ such that

$$\begin{aligned}
&\boldsymbol{A}(s)\partial_t^2\boldsymbol{r}(s,t) - \partial_s\big(\boldsymbol{B}(s,t)\partial_s\boldsymbol{r}(s,t)\big) = \boldsymbol{C}(s,t) && \forall \quad (s,t)\in\Omega\,,\\
&\boldsymbol{B}(0,t)\partial_s\boldsymbol{r}(0,t) = \boldsymbol{f}(t),\ \boldsymbol{r}(L,t) = \boldsymbol{\gamma}(t),\ \boldsymbol{B}(L,t)\partial_s\boldsymbol{r}(L,t) = \boldsymbol{\eta}(t) && \forall \quad t\in\mathcal{T}\,,\\
&\boldsymbol{f}(0) = \boldsymbol{f}_0 \quad\text{and}\quad \boldsymbol{r}(s,0) = \boldsymbol{r}_0(s),\ \partial_t\boldsymbol{r}(s,0) = \boldsymbol{v}_0(s) && \forall \quad s\in S\,,
\end{aligned} \tag{2.11}$$

where, the abbreviations $\boldsymbol{A} : S \mapsto \mathbb{R}^{d,d}$, $\boldsymbol{B} : \Omega \mapsto \mathbb{R}^{d,d}$ and $\boldsymbol{C} : \Omega \mapsto \mathbb{R}^{d}$, as given by

$$\boldsymbol{A}(s) = (\rho A)(s)\boldsymbol{I}_d, \qquad \boldsymbol{B}(s,t) = \frac{N(s,t)}{\nu(s,t)}\boldsymbol{I}_d \quad\text{and}\quad \boldsymbol{C}(s,t) = \boldsymbol{b}(s,t) \tag{2.12}$$

have been introduced. Note that $\boldsymbol{I}_d$ denotes the $d\times d$ identity matrix. The system of partial differential equations $(2.11)_1$ results from substituting (2.3) into the equations of motion (2.1). Note that $(2.11)_1$ can be classified as a second-order quasilinear hyperbolic system of partial differential equations in one dimension for $\boldsymbol{r}(s,t)$ (cf. [7]). The nonlinearity of $(2.11)_1$ is reflected in the dependence of $\boldsymbol{B}(s,t)$ on the unknown vector $\boldsymbol{r}(s,t)$, cf. $(2.12)_2$ together with (2.4) and (2.5).

Moreover, $(2.11)_2$ corresponds to the boundary conditions (2.10) and (2.9), respectively. Accordingly, if the end at $s = L$ of the string is free, we have $\boldsymbol{\eta}(t) = \boldsymbol{0}$. Note, however, that the force vector $\boldsymbol{\eta}(t)$ has been introduced to take into account more general scenarios such as those mentioned in Remarks 2.2 and 2.3.

Initial conditions are accounted for by $(2.11)_3$, where $\boldsymbol{f}_0 \in \mathbb{R}^d$ and $\boldsymbol{r}_0, \boldsymbol{v}_0 \in \mathbb{R}^d$ are prescribed. Feedforward control problems often deal with rest-to-rest maneuvers. There, the solution of the corresponding equilibrium problem supplies the initial values $\boldsymbol{f}_0$, $\boldsymbol{r}_0$ and $\boldsymbol{v}_0 = \boldsymbol{0}$. In the more general case in which the initial configuration of the string is not at rest, the initial values for the inverse problem can be obtained by solving the corresponding forward dynamics problem.

Remark 2.2 (Additional point mass). *An additional point mass M might be attached to the boundary of the string at $s = L$. In this case, the last boundary condition in* $(2.11)_2$ *would have to be replaced by*

$$\boldsymbol{\eta}(t) = M\left(\boldsymbol{g} - \partial_t^2\boldsymbol{r}(L,t)\right) \tag{2.13}$$

where $\partial_t^2\boldsymbol{r}(L,t) = \mathrm{D}_t^2\boldsymbol{\gamma}(t)$ and, as before, $\boldsymbol{g}$ denotes the vector of gravitational acceleration.

Remark 2.3 (Multibody system). *If the elastic string is a sub-system belonging to the inverse dynamics problem of a multibody system, the force vector $\boldsymbol{\eta}(t)$ in the last boundary condition in* $(2.11)_2$ *is a prescribed function of time (see Chap. 5 for further details).*

Remark 2.4. *In the context of infinite-dimensional models for strings, flatness-based methods have been proposed in [55, 60, 76] to solve the inverse dynamics problem (see Def. 3.23 and accompanying examples and references therein for more background on differentially flat systems). The flatness-based methods typically rely on mechanical modeling assumptions such as small deformations [55, 76] or inextensibility [60].*

Remark 2.5 (Linear-elastic bar). *Concerning small longitudinal deformations about a straight reference configuration, we recover the model of a linear-elastic bar (Fig. 2.3).*

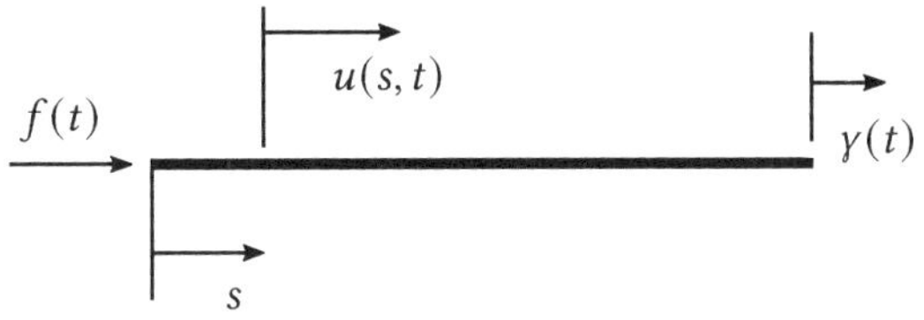

Figure 2.3.: Illustration of the linear-elastic bar.

Accordingly, consider the one-parameter family of configurations $\mathbf{r}^\varepsilon(s,t) = r^\varepsilon(s,t)\mathbf{e}$, *where* $\mathbf{e}$ *is a unit vector and* r^ε *is a power series expansion of the form*

$$r^\varepsilon(s,t) = s + \varepsilon u(s,t) + \mathcal{O}(\varepsilon^2) \ .$$

Here, $r^0 = s$ *specifies the placement of material points in the straight reference configuration and the derivative* $\frac{d}{d\varepsilon}\big|_{\varepsilon=0} r^\varepsilon(s,t) = u(s,t)$ *characterizes small longitudinal displacements. Moreover, the resulting axial strains can be obtained via*

$$\epsilon(s,t) = \frac{d}{d\varepsilon}\bigg|_{\varepsilon=0} \partial_s r^\varepsilon(s,t) = \partial_s u(s,t) \ .$$

Similarly, the linearized stretch yields $\nu = 1 + \partial_s u$, *so that the linearized constitutive relation* (2.8) *can be written as* $\hat{N}_{\text{lin}} = EA\epsilon$. *Eventually, linearizing the equations of motion* $(2.11)_1$ *of the string about the straight reference configuration and neglecting the body forces yields*

$$\rho A \partial_{tt}^2 u - EA \partial_{ss}^2 u = 0 \ . \tag{2.14}$$

This partial differential equation governs the longitudinal motion of the linear-elastic bar and coincides with the wave equation in one dimension. The inverse dynamics problem (2.11) *can now be restated for the linear-elastic bar: Find* $u : \Omega \mapsto \mathbb{R}$ *and* $f : \mathcal{T} \mapsto \mathbb{R}$ *such that*

$$\begin{aligned} &\partial_{tt}^2 u - c^2 \partial_{ss}^2 u = 0 && \forall \quad (s,t) \in \Omega \ , \\ &EA\partial_s u(0,t) = f(t), \ u(L,t) = \gamma(t), \ \partial_s u(L,t) = 0 && \forall \quad t \in \mathcal{T} \ , \\ &r(s,0) = s, \ \partial_t r(s,0) = 0 && \forall \quad s \in \mathcal{S} \ . \end{aligned} \tag{2.15}$$

Here, for the sake of clarity and without loss of generality, a free end at $s = L$ *is assumed. In* $(2.15)_1$, c *denotes the constant wave propagation speed defined by*

$$c = \sqrt{\frac{E}{\rho}} \ . \tag{2.16}$$

There exists an analytical solution to problem (2.15) *that will serve as reference for the numerical methods developed in the sequel.*

Remark 2.6 (Analytical solution to Ex. 2.5). *An analytical solution to problem* (2.15) *is based on d'Alembert's formula (cf. e.g. [39] or [36, Ch. 16]). It can be easily verified that the ansatz*

$$u(s,t) = \Phi(t + c^{-1}s) + \Psi(t - c^{-1}s) \tag{2.17}$$

satisfies the partial differential equation $(2.15)_1$. *Now, the first boundary condition in* $(2.15)_2$ *yields* $f(t) = EA\partial_s u(0,t)$, *where* (2.17) *implies*

$$\partial_s u(0,t) = (\Phi'(t) - \Psi'(t))\, c^{-1} \ . \tag{2.18}$$

Inserting (2.17) *into the second boundary condition,* $u(L,t) = \gamma(t)$, *and subsequently differentiating with respect to time yields*

$$\Phi'(t + c^{-1}L) + \Psi'(t - c^{-1}L) = \gamma'(t) \ . \tag{2.19}$$

Similarly, the third boundary condition, i.e. $\partial_s u(L,t) = 0$, *gives rise to*

$$\Phi'(t + c^{-1}L) - \Psi'(t - c^{-1}L) = 0 \ . \tag{2.20}$$

Adding (2.19) *and* (2.20) *yields* $2\Phi'(t + c^{-1}L) = \gamma'(t)$, *while subtracting* (2.20) *from* (2.19) *yields* $2\Psi'(t - c^{-1}L) = \gamma'(t)$. *Evaluating the last two equations at* $\bar{t} = t - c^{-1}L$ *and* $\bar{t} = t + c^{-1}L$, *respectively, leads to*

$$\begin{aligned} \Phi'(t) &= \frac{1}{2}\gamma'(t - c^{-1}L) \ , \\ \Psi'(t) &= \frac{1}{2}\gamma'(t + c^{-1}L) \ . \end{aligned} \tag{2.21}$$

Inserting (2.21) *into* (2.18), *the first boundary condition eventually yields the result*

$$f(t) = \frac{EA}{2c}\left(\gamma'(t - c^{-1}L) - \gamma'(t + c^{-1}L)\right) \ , \tag{2.22}$$

providing the actuating force at the left boundary of the bar in terms of the prescribed displacement of the right boundary of the bar.

Remark 2.7 (Analytical solution to Ex. 2.5 - alternative). *Considering the one dimensional linear wave equation* $(2.15)_1$ *governing the motion of linear elastic bars, a Laplace transformation eventually gives rise to an analytical solution as well. Therefore, the following second-order ordinary differential equation with constant coefficients*

$$\tau^2 U(s,\tau) - c^2 \partial^2_{ss} U(s,\tau) = 0 \tag{2.23}$$

subjected to the following boundary conditions

$$EA\partial_s U(0,\tau) - F(\tau) = 0, \qquad EA\partial_s U(L,\tau) = 0 \quad \text{and} \qquad U(L,\tau) - \Gamma(\tau) = 0 \tag{2.24}$$

can be obtained. In Eq. (2.24), $\Gamma(\tau)$ *and* $F(\tau)$ *denote the Laplace transforms of the trajectory and actuating force, respectively. Consulting any literature on ordinary differential equations (e.g. [92]), the exponential ansatz*

$$U(s,\tau) = A(\tau)\exp(\lambda s) \ , \tag{2.25}$$

solves the differential equation (2.23) *(e.g. [92]). Inserting* (2.25) *as well as its first and second derivative*

$$\partial_s U(s,\tau) = \lambda A(\tau)\exp(\lambda s) \quad \text{and} \quad \partial^2_{ss} U(s,\tau) = \lambda^2 A(\tau)\exp(\lambda s)$$

into (2.23) *yields*

$$\tau^2 A(\tau)\exp(\lambda s) - c^2\lambda^2 A(\tau)\exp(\lambda s) = 0\ . \tag{2.26}$$

Since $\lambda_1 = -\tau c^{-1}$ *and* $\lambda_2 = \tau c^{-1}$, *solves* (2.26), *it applies that*

$$U(s,\tau) = A_1(\tau)\exp(-\tau s c^{-1}) + A_2(\tau)\exp(\tau s c^{-1}) \tag{2.27}$$

constitutes a fundamental solution to (2.23). *Taking the boundary conditions* (2.24) *into account, it follows*

$$\begin{aligned}
U(s=L,\tau) &= A_1(\tau)\exp(-\tau L c^{-1}) + A_2(\tau)\exp(\tau L c^{-1}) = \Gamma(\tau)\ ,\\
EA\partial_s U(L,\tau) &= EA\left(-c^{-1}\tau A_1(\tau)\exp(-\tau L c^{-1}) + c^{-1}\tau A_2(\tau)\exp(\tau L c^{-1})\right) = 0\ ,\\
EA\partial_s U(0,\tau) &= EA\left(-c^{-1}\tau A_1(\tau) + c^{-1}\tau A_2(\tau)\right) = F(\tau)\ .
\end{aligned}$$

The unknown functions $A_1(\tau)$, $A_2(\tau)$ *and* $F(\tau)$ *are uniquely determined as*

$$\begin{aligned}
A_1(\tau) &= \frac{1}{2}\exp(\tau L c^{-1})\Gamma(\tau)\ ,\\
A_2(\tau) &= \frac{1}{2}\exp(-\tau L c^{-1})\Gamma(\tau)\ ,\\
F(\tau) &= \frac{EA}{2}c^{-1}\tau\Gamma(\tau)\left(\exp(-\tau L c^{-1}) - \exp(\tau L c^{-1})\right)\ .
\end{aligned} \tag{2.28}$$

Inserting $(2.28)_1$ *and* $(2.28)_2$ *into* (2.27) *leads to*

$$U(s,\tau) = \frac{1}{2}\Gamma(\tau)\left(\exp(-\left(c^{-1}(x-L)\right)\tau) + \exp(-\left(c^{-1}(L-x)\right)\tau)\right) \tag{2.29}$$

and consequently

$$\partial_s U(s,\tau) = -c^{-1}\tau A_1(\tau)\exp(-c^{-1}\tau s) + c^{-1}\tau A_2(\tau)\exp(c^{-1}\tau s)\ . \tag{2.30}$$

Transforming back into the time domain, by taking advantage of the shifting rule leads to the solution

$$u(s,t) = \frac{1}{2}\gamma(t - c^{-1}(s-L)) + \frac{1}{2}\gamma(t - c^{-1}(L-s))\ . \tag{2.31}$$

The actuating force can then be determined as (cf. (2.22)*)*

$$f(t) = \frac{EA}{2c}\left(\gamma'(t - c^{-1}L) - \gamma'(t + c^{-1}L)\right)\ . \tag{2.32}$$

2.2. Geometrically exact beam formulation

This section aims to briefly derive the *classical equations of motion of geometrically exact beams*. Furthermore we will demonstrate, that these equations align with the framework proposed in Sec. 1.1. Essentially, the derivations, that are made subsequently, are based on [7], [88], [77] and [20].

Kinematics. The motion of any material point $s \in S = [0, L] \subset \mathbb{R}$ of a beam at any time $t \in \mathcal{T} = [0, \infty) \subset \mathbb{R}$ is defined by the deformation map

$$\boldsymbol{r} : \Omega \mapsto \mathbb{R}^3 \tag{2.33}$$

determining the configuration of a curve[2], along with a moving orthonormal basis

$$\boldsymbol{d}_i : \Omega \mapsto \mathbb{R}^3 \qquad \forall i \in \{1, 2, 3\} \tag{2.34}$$

indicating the *average orientation of the cross-section*, where $\boldsymbol{d}_3$ is normal and $\boldsymbol{d}_1$ and $\boldsymbol{d}_2$ are tangential to the cross-section. In the reference configuration, the beam is defined by

$$\boldsymbol{r}|_{(s,t)\in\partial\Omega_0} \equiv \boldsymbol{R}(s) : \partial\Omega_0 \mapsto \mathbb{R}^3 \qquad \text{and} \qquad \boldsymbol{d}_i|_{(s,t)\in\partial\Omega_0} \equiv \boldsymbol{D}_i(s) : \partial\Omega_0 \mapsto \mathbb{R}^3$$

for $\partial\Omega_0 = S \times \{0\}$. By introducing the proper-orthogonal tensor $\boldsymbol{\Lambda} \in SO(3)$, the rotation of the orthonormal basis (2.34) is given by

$$\boldsymbol{d}_i = \boldsymbol{\Lambda}\boldsymbol{D}_i. \tag{2.35}$$

At this point, it should be emphasized that planarity of the cross-section is assumed. Abandoning this, would require further spatial variables.

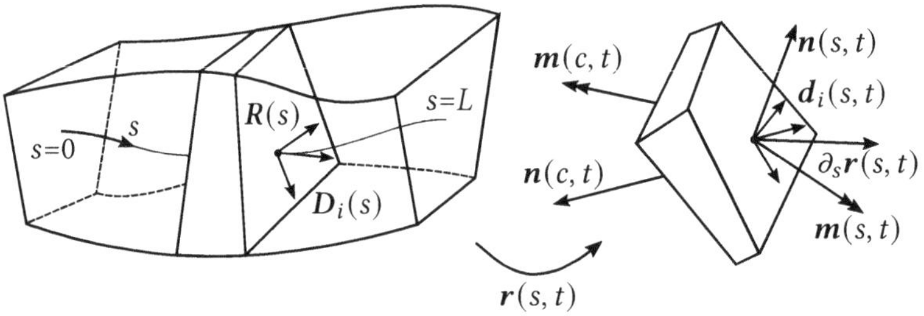

Figure 2.4.: Illustration of the kinematics of the geometrically exact beam formulation.

Differentiating the orthonormal basis (2.34) with respect to the spatial variable s yields

$$\partial_s \boldsymbol{d}_i = (\partial_s \boldsymbol{\Lambda})\boldsymbol{D}_i + \boldsymbol{\Lambda}\partial_s \boldsymbol{D}_i = (\partial_s \boldsymbol{\Lambda})\boldsymbol{\Lambda}^T \boldsymbol{d}_i + \boldsymbol{\Lambda}(\partial_s \boldsymbol{\Lambda}_0)\boldsymbol{\Lambda}_0^T \boldsymbol{\Lambda}^T \boldsymbol{d}_i \ . \tag{2.36}$$

[2] This axis must not necessarily coincide with the line of centroids of the beam

In Eq. (2.36), the proper-orthogonal tensor $\boldsymbol{\Lambda}_0 \in SO(3)$ has been introduced determining the shape of the beam in reference configuration, i.e. $\boldsymbol{D}_i = \boldsymbol{\Lambda}_0 \boldsymbol{e}_i$ and consequently

$$\partial_s \boldsymbol{D}_i = \partial_s(\boldsymbol{\Lambda}_0 \boldsymbol{e}_i) = (\partial_s \boldsymbol{\Lambda}_0)\boldsymbol{e}_i + \boldsymbol{\Lambda}_0 \partial_s \boldsymbol{e}_i = (\partial_s \boldsymbol{\Lambda}_0)\boldsymbol{\Lambda}_0^T \boldsymbol{D}_i = (\partial_s \boldsymbol{\Lambda}_0)\boldsymbol{\Lambda}_0^T \boldsymbol{\Lambda}^T \boldsymbol{d}_i \ .$$

By introducing the skew-symmetric curvature matrix[3]

$$\boldsymbol{S}_{\boldsymbol{\kappa}} = (\partial_s \boldsymbol{\Lambda})\boldsymbol{\Lambda}^T = \begin{bmatrix} 0 & -\kappa_3 & \kappa_2 \\ \kappa_3 & 0 & -\kappa_1 \\ -\kappa_2 & \kappa_1 & 0 \end{bmatrix} \tag{2.37}$$

Eq. (2.36) can be rewritten as

$$\partial_s \boldsymbol{d}_i = (\boldsymbol{S}_{\boldsymbol{\kappa}} + \boldsymbol{\Lambda} \boldsymbol{S}_{\boldsymbol{\kappa}}^0 \boldsymbol{\Lambda}^T)\boldsymbol{d}_i = \bar{\boldsymbol{S}}_{\boldsymbol{\kappa}} \boldsymbol{d}_i \ . \tag{2.38}$$

Note that the assumption of a straight reference configuration implies $\boldsymbol{\Lambda}_0 \equiv \boldsymbol{I}$ and hence $\boldsymbol{S}_{\kappa}^0 \equiv \boldsymbol{0}$. Introducing the axial vector $\boldsymbol{\kappa} = \kappa_i \boldsymbol{d}_i$, Eq. (2.38) can be defined alternatively as

$$\partial_s \boldsymbol{d}_i = \boldsymbol{\kappa} \times \boldsymbol{d}_i \ .$$

Analogously, the temporal change of the moving basis can be obtained by introducing $\boldsymbol{\omega} = \omega_i \boldsymbol{d}_i$, as

$$\partial_t \boldsymbol{d}_i = \boldsymbol{S}_{\omega} \boldsymbol{d}_i = \boldsymbol{\omega} \times \boldsymbol{d}_i \ .$$

Herein $\boldsymbol{S}_{\omega}$ represents the skew-symmetric angular velocity matrix. Note, that by definition $\boldsymbol{S}_{\omega}\boldsymbol{\omega} \equiv \boldsymbol{0}$. Following [7, pp. 260-261], the strain variables are defined as

$$\gamma_i = \partial_s \boldsymbol{r} \cdot \boldsymbol{d}_i \quad \text{and} \quad \kappa_i = \boldsymbol{\kappa} \cdot \boldsymbol{d}_i \ , \tag{2.39}$$

where γ_1 and γ_2 measure shear, γ_3 measures dilatation, κ_1 and κ_2 measure flexure and κ_3 measures torsion.

Example 2.8 (Planar kinematics). *Regarding planar motion, the rotation around the director $\boldsymbol{d}_2(s,t) \equiv \boldsymbol{D}_2(s)$ for all $t \in T$ can be described by the following proper-orthogonal matrix*

$$\boldsymbol{\Lambda} = \begin{bmatrix} \cos\Theta & 0 & -\sin\Theta \\ 0 & 1 & 0 \\ \sin\Theta & 0 & \cos\Theta \end{bmatrix} . \tag{2.40}$$

With (2.40) *the angular velocity matrix $\boldsymbol{S}_{\omega} = \partial_t(\boldsymbol{\Lambda})\boldsymbol{\Lambda}^T$ simplifies to*

$$\boldsymbol{S}_{\omega} = \partial_t \Theta \begin{bmatrix} -\sin\Theta & 0 & -\cos\Theta \\ 0 & 0 & 0 \\ \cos\Theta & 0 & -\sin\Theta \end{bmatrix} \begin{bmatrix} \cos\Theta & 0 & \sin\Theta \\ 0 & 1 & 0 \\ -\sin\Theta & 0 & \cos\Theta \end{bmatrix} = \partial_t \Theta \begin{bmatrix} 0 & 0 & -1 \\ 0 & 0 & 0 \\ 1 & 0 & 0 \end{bmatrix} . \tag{2.41}$$

[3] Orthogonality of $\boldsymbol{\Lambda}$ implies $\boldsymbol{\Lambda}\boldsymbol{\Lambda}^T = \boldsymbol{I}$. Differentiation of this orthogonality condition with respect to the spatial variable s yields $\partial_s(\boldsymbol{\Lambda})\boldsymbol{\Lambda}^T + \boldsymbol{\Lambda}\partial_s(\boldsymbol{\Lambda}^T) = 0$ which indicates the skew-symmetry of $\partial_s(\boldsymbol{\Lambda})\boldsymbol{\Lambda}^T$

Consequently the axial vector is given by $\boldsymbol{\omega} = \omega_2 \boldsymbol{d}_2 = \omega_2 \boldsymbol{e}_2$. *The same can be shown for the curvature* $\boldsymbol{\kappa} = \kappa_2 \boldsymbol{d}_2 = \kappa_2 \boldsymbol{e}_2$ *by computing the curvature matrix in* (2.37) *using* (2.40).

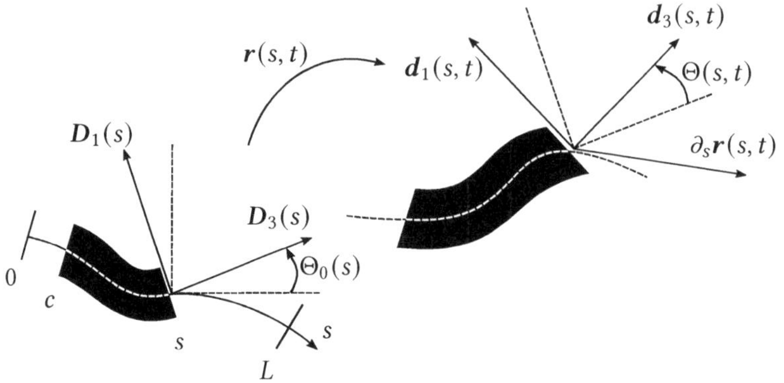

Figure 2.5.: Planar illustration of the kinematics of the geometrically exact beam formulation.

Equilibrium. After having addressed the kinematics, the corresponding dynamics will be investigated subsequently. The (material form of the) balance of linear momentum on an interval $S \supset \mathcal{I} = [c, s]$ of a beam can be established as follows

$$\boldsymbol{n}(s,t) - \boldsymbol{n}(c,t) + \int_c^s \bar{\boldsymbol{n}}(\xi_3) \; \mathrm{d}\xi_3 = \partial_t \boldsymbol{P}(s,t) \; . \tag{2.42}$$

Therein, the contact force $\boldsymbol{n} : \Omega \mapsto \mathbb{R}^3$, the external load $\bar{\boldsymbol{n}} : \Omega \mapsto \mathbb{R}^3$ and the linear momentum $\boldsymbol{P} : \Omega \mapsto \mathbb{R}^3$ acting on a beam segment $\mathcal{I}$ have been introduced.

Since the center of gravity of each cross-section is specified as $\boldsymbol{r}_S = \boldsymbol{r} + \xi_\alpha^S \boldsymbol{d}_\alpha$ for $\xi_\alpha^S = A^{-1} \int \xi_\alpha \; \mathrm{d}A = A^{-1} S_\alpha$, where $S_\alpha : S \mapsto \mathbb{R} \forall \alpha \in \{1,2\}$ is the first moment of area with respect to $\xi_\alpha \in \mathbb{R}$ (cf. Fig. 2.6), the linear momentum $\boldsymbol{P} : \Omega \mapsto \mathbb{R}^3$ can be obtained as

$$\begin{aligned}
\boldsymbol{P}(s,t) &= \int \partial_t \boldsymbol{r}_S \; \mathrm{d}m \\
&= \int_c^s (\rho A)(s) \partial_t (\boldsymbol{r} + \xi_\alpha^S \boldsymbol{d}_\alpha) \; \mathrm{d}\xi_3 \\
&= \int_c^s (\rho A)(s) \partial_t \boldsymbol{r} \; \mathrm{d}\xi + \int_c^s \rho(s) \int \xi_\alpha \, \mathrm{d}A \; \partial_t \boldsymbol{d}_\alpha \; \mathrm{d}\xi_3 \\
&= \int_c^s (\rho A)(s) \partial_t \boldsymbol{r} + (\rho S_\alpha)(s) \partial_t \boldsymbol{d}_\alpha \; \mathrm{d}\xi_3 \qquad \forall \alpha \in \{1,2\} \\
&= \int_c^s \boldsymbol{p}(s,t) \; \mathrm{d}\xi_3 \; .
\end{aligned}$$

Therein, the linear momentum density function $\boldsymbol{p} : \Omega \mapsto \mathbb{R}^3$ as given by

$$\boldsymbol{p} = (\rho A)(s) \partial_t \boldsymbol{r} + (\rho S_\alpha)(s) \partial_t \boldsymbol{d}_\alpha \tag{2.43}$$

has been introduced. Defining the linear momentum density relative to $\boldsymbol{r}(s,t)$ as

$$\partial_t \boldsymbol{q}(s,t) \equiv (\rho S_\alpha)(s)\partial_t \boldsymbol{d}_\alpha \qquad \forall \alpha \in \{1,2\}\,, \tag{2.44}$$

the derivative of the linear momentum density $\boldsymbol{p}(s,t)$ with respect to time is obtained as

$$\partial_t \boldsymbol{p}(s,t) = (\rho A)(s)\partial_t^2 \boldsymbol{r} + (\rho S_\alpha)(s)\partial_t^2 \boldsymbol{d}_\alpha = (\rho A)(s)\partial_t^2 \boldsymbol{r} + \partial_t^2 \boldsymbol{q} \qquad \forall \alpha \in \{1,2\}\,. \tag{2.45}$$

Obviously, the relative linear momentum density function $\boldsymbol{q} : \Omega \mapsto \mathbb{R}$, defined in (2.44), vanishes by choosing $\boldsymbol{r}(s,t)$ accurately, i.e. $\boldsymbol{r}(s,t) = \boldsymbol{r}_S(s,t)$, such that $S_\alpha = 0$ for $\alpha \in \{1,2\}$. Analogously, the material form of the balance of angular momentum for a segment $\mathcal{I}$ can be established as

$$\begin{aligned}\boldsymbol{m}(s,t) - \boldsymbol{m}(c,t) + (\boldsymbol{r}(s,t) \times \boldsymbol{n}(s,t)) - (\boldsymbol{r}(c,t) \times \boldsymbol{n}(c,t))\\ + \int_c^s \boldsymbol{r}(\xi) \times \bar{\boldsymbol{n}}(\xi)\ \mathrm{d}\xi + \int_c^s \bar{\boldsymbol{m}}(\xi)\ \mathrm{d}\xi = \partial_t \boldsymbol{L}(s,t)\,.\end{aligned} \tag{2.46}$$

Therein, the contact torque $\boldsymbol{m} : \Omega \mapsto \mathbb{R}^3$, the external applied torque $\bar{\boldsymbol{m}} : \Omega \mapsto \mathbb{R}^3$ and the angular momentum (relative to an inertial frame) $\boldsymbol{L} : \Omega \mapsto \mathbb{R}^3$ of a beam segment $\mathcal{I}$ have been introduced. Taking advantage of the definition of the second moment of area $I_{\alpha\beta} = \int \xi_\alpha \xi_\beta\ \mathrm{d}A$ and assuming $\boldsymbol{r}_P = \boldsymbol{r} + \xi_\alpha \boldsymbol{d}_\alpha$ (cf. Fig. 2.6), the angular momentum $\boldsymbol{L}(s,t)$ can be obtained as

$$\begin{aligned}\boldsymbol{L}(s,t) &= \int \boldsymbol{r}_P \times \partial_t(\boldsymbol{r}_P)\,\mathrm{d}m\\ &= \int (\boldsymbol{r} + \xi_\alpha \boldsymbol{d}_\alpha) \times (\partial_t \boldsymbol{r} + \xi_\beta \partial_t \boldsymbol{d}_\beta)\,\mathrm{d}m\\ &= \rho \int_c^s \int \boldsymbol{r} \times \partial_t \boldsymbol{r}\,\mathrm{d}A + \int \xi_\beta(\boldsymbol{r} \times \partial_t \boldsymbol{d}_\beta) + \xi_\alpha(\boldsymbol{d}_\alpha \times \partial_t \boldsymbol{r})\,\mathrm{d}A\\ &\qquad + \int \xi_\alpha \xi_\beta(\boldsymbol{d}_\alpha \times \partial_t \boldsymbol{d}_\beta)\,\mathrm{d}A\,\mathrm{d}\xi_3\\ &= \int_c^s \rho A(\boldsymbol{r} \times \partial_t \boldsymbol{r}) + \rho S_\alpha(\boldsymbol{r} \times \partial_t \boldsymbol{d}_\alpha + \boldsymbol{d}_\alpha \times \partial_t \boldsymbol{r}) + \rho I_{\alpha\beta}(\boldsymbol{d}_\alpha \times \partial_t \boldsymbol{d}_\beta)\ \mathrm{d}\xi_3\\ &= \int_c^s \boldsymbol{l}(s,t)\ \mathrm{d}\xi_3 \quad \forall \alpha,\beta \in \{1,2\}\,,\end{aligned} \tag{2.47}$$

where the angular momentum density function $\boldsymbol{l} : \Omega \mapsto \mathbb{R}^3$ as given by

$$\boldsymbol{l}(s,t) = \rho A(\boldsymbol{r} \times \partial_t \boldsymbol{r}) + \rho S_\alpha(\boldsymbol{r} \times \partial_t \boldsymbol{d}_\alpha + \boldsymbol{d}_\alpha \times \partial_t \boldsymbol{r}) + \rho I_\alpha(\boldsymbol{d}_\alpha \times \partial_t \boldsymbol{d}_\alpha) \quad \forall \alpha,\beta \in \{1,2\} \tag{2.48}$$

has been introduced. By defining the angular momentum relative to $\boldsymbol{r}(s,t)$ as

$$\boldsymbol{h}(s,t) \equiv \rho I_{\alpha\beta}(\boldsymbol{d}_\alpha \times \partial_t \boldsymbol{d}_\beta) \qquad \forall \alpha,\beta \in \{1,2\} \tag{2.49}$$

the time derivative of the angular momentum density function can be established as

$$\begin{aligned} \partial_t \boldsymbol{l}(s,t) &= \rho A(\boldsymbol{r} \times \partial_t^2 \boldsymbol{r}) + \rho S_\alpha(\boldsymbol{d}_\alpha \times \partial_t^2 \boldsymbol{r}) + \rho S_\alpha(\boldsymbol{r} \times \partial_t^2 \boldsymbol{d}_\alpha) + \rho I_{\alpha\beta}(\boldsymbol{d}_\alpha \times \partial_t^2 \boldsymbol{d}_\beta) \\ &= \rho A(\boldsymbol{r} \times \partial_t^2 \boldsymbol{r}) + \boldsymbol{r} \times \partial_t^2 \boldsymbol{q} + \boldsymbol{q} \times \partial_t^2 \boldsymbol{r} + \partial_t \boldsymbol{h} \,. \end{aligned} \tag{2.50}$$

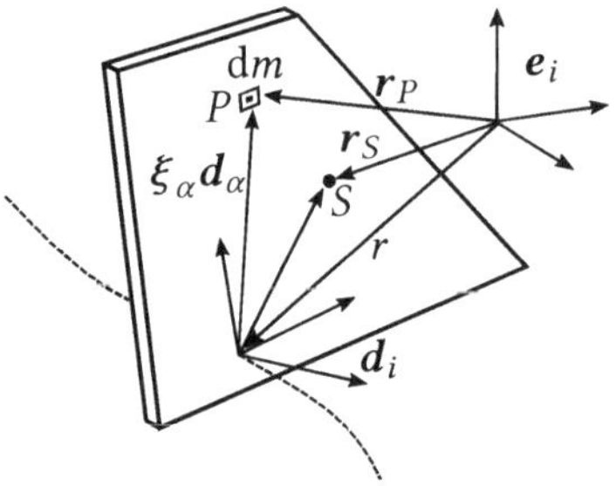

Figure 2.6.: Illustration of the geometry of the cross-section of the geometrically exact beam.

Taking the derivative of (2.42) and (2.46) with respect to the spatial variable $s \in S$, the balance equations are represented in local form as given by

$$\begin{aligned} \partial_s \boldsymbol{n} + \bar{\boldsymbol{n}} &= \partial_t \boldsymbol{p} \\ \partial_s \boldsymbol{m} + (\boldsymbol{r} \times \partial_s \boldsymbol{n}) + (\partial_s \boldsymbol{r} \times \boldsymbol{n}) + (\boldsymbol{r} \times \bar{\boldsymbol{n}}) + \bar{\boldsymbol{m}} &= \partial_t \boldsymbol{l}. \end{aligned} \tag{2.51}$$

Inserting Eq. (2.45) into Eq. $(2.51)_1$, the following relation can be established

$$\boldsymbol{r} \times \partial_t \boldsymbol{p} = \rho A(\boldsymbol{r} \times \partial_t^2 \boldsymbol{r}) + \boldsymbol{r} \times \partial_t^2 \boldsymbol{q} = \boldsymbol{r} \times \partial_s \boldsymbol{n} + \boldsymbol{r} \times \bar{\boldsymbol{n}}.$$

By introducing $\partial_t \hat{\boldsymbol{l}} \equiv \boldsymbol{q} \times \partial_t^2 \boldsymbol{r} + \partial_t \boldsymbol{h}$, the balance of angular momentum in local form $(2.51)_2$ might be reformulated as

$$\partial_s \boldsymbol{m} + \partial_s \boldsymbol{r} \times \boldsymbol{n} + \bar{\boldsymbol{m}} = \partial_t \hat{\boldsymbol{l}}, \tag{2.52}$$

Equation $(2.51)_1$ and (2.52) are the *equations of motion for (the classical form of Cosserat) rods* ([30], [7]). These equations are also frequently and synonymously referred to as the *geometrically exact* ([20]) or the *Simo-Reissner* beam formulation ([88],[89], [77] and [78]). It might be convenient to write them compactly as

$$\begin{bmatrix} \boldsymbol{p} \\ \hat{\boldsymbol{l}} \end{bmatrix}_{,t} - \begin{bmatrix} \boldsymbol{n} \\ \boldsymbol{m} \end{bmatrix}_{,s} = \begin{bmatrix} 0 \\ \partial_s \boldsymbol{r} \times \boldsymbol{n} \end{bmatrix} + \begin{bmatrix} \bar{\boldsymbol{n}} \\ \bar{\boldsymbol{m}} \end{bmatrix}. \tag{2.53}$$

Example 2.9 (Planar equilibrium)**.** *For the plane case, the angular momentum relative to $\boldsymbol{r}(s,t)$, defined in* (2.49)*, and its time derivative reduces to*

$$\boldsymbol{h}(s,t) = \rho I_{11}(\boldsymbol{d}_1 \times \partial_t \boldsymbol{d}_1) = \omega_2 \boldsymbol{e}_2 \quad \textit{and} \quad \partial_t \boldsymbol{h}(s,t) = (\partial_t \omega_2) \boldsymbol{e}_2,$$

respectively, due to $\partial_t \boldsymbol{d}_2 = \partial_t^2 \boldsymbol{d}_2 = 0$ *and*

$$\boldsymbol{d}_1 \times \partial_t \boldsymbol{d}_1 = \boldsymbol{\Lambda} \boldsymbol{e}_1 \times \boldsymbol{S}_\omega \boldsymbol{\Lambda} \boldsymbol{e}_1 = \omega_2 \boldsymbol{e}_2. \tag{2.54}$$

Therein use of the kinematical relations, presented in Ex. 2.8 for the plane case, has been made. In Fig. 2.7, the geometry of a cross-section for the planar case is depicted.

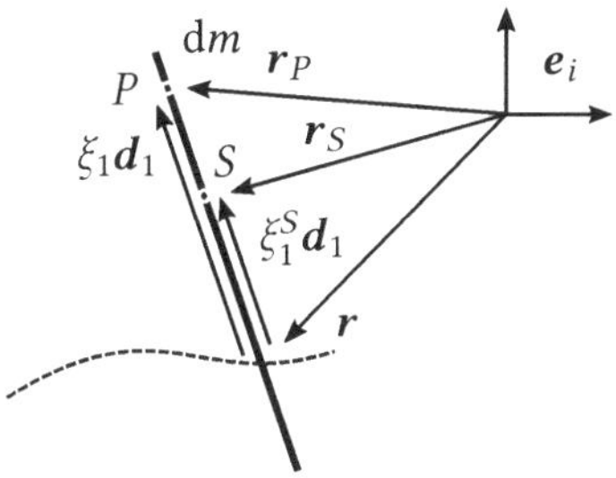

Figure 2.7.: Planar illustration of the geometry of the cross-section of the geometrically exact beam.

Furthermore, the contact force and contact moment are assumed as $\boldsymbol{n} \equiv (n_1, 0, n_3)^T$ *and* $\boldsymbol{m} \equiv (0, m_2, 0)^T$*, respectively (cf. Fig. 2.8). The same holds for the external applied force and moment* $\bar{\boldsymbol{n}}$ *and* $\bar{\boldsymbol{m}}$*, respectively.*

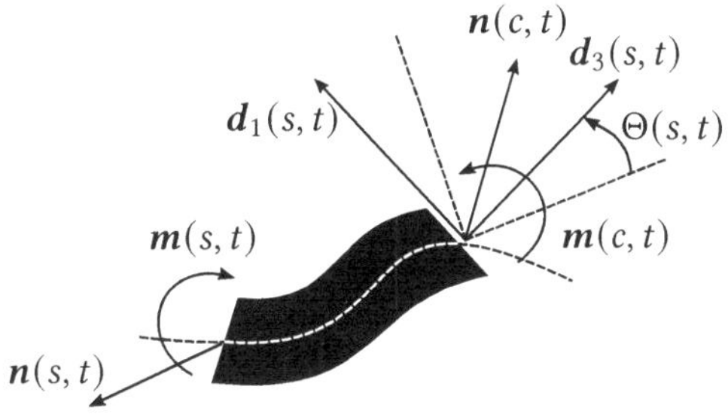

Figure 2.8.: Planar illustration of the equillibrium of the geometrically exact beam formulation.

Subsequently, it will be demonstrated, that the classical equations of motion for *Cosserat* rods, introduced in $(2.51)_1$ and (2.52), align with the mathematical framework postulated in Sec. 1.1. For this, the contact force and moment can be written alternatively as

$$\boldsymbol{n} = N_i \boldsymbol{d}_i = N_i \boldsymbol{\Lambda} \boldsymbol{e}_i \quad \text{and} \quad \boldsymbol{m} = M_i \boldsymbol{d}_i = M_i \boldsymbol{\Lambda} \boldsymbol{e}_i \ .$$

Additionally we need to resolve the geometric constraints, that are imposed on the components of the directors accounting for the orthonormality. One possible way of doing this is by introducing *Euler angles* $\boldsymbol{\Theta} : \Omega \mapsto \mathbb{R}^3$. The *Euler angles* form a local basis in $SO(3)$ such that the overall rotation $\boldsymbol{\Lambda}(\boldsymbol{\Theta}) \in SO(3)$ of a moving frame $\{\boldsymbol{d}_i\}$ relative to an inertial frame is defined by *three* rotations that are carried out successively: (i) rotate by

Θ_2 around $\boldsymbol{e}_3$, (ii) rotate by Θ_1 around $\boldsymbol{h}$ and (iii) rotate by Θ_3 around $\boldsymbol{d}_3$. See Fig. 2.9 for a graphical illustration of these rotations.

Following [7, pp. 300-301], the strain variables are then given as

$$\kappa_k = (\boldsymbol{R}\partial_s\boldsymbol{\Theta})_k \quad \text{and} \quad \omega_k = (\boldsymbol{R}\partial_t\boldsymbol{\Theta})_k \tag{2.55}$$

where

$$\boldsymbol{R} = \begin{bmatrix} \sin\Theta_3 & -\cos\Theta_3\sin\Theta_1 & 0 \\ \cos\Theta_3 & \sin\Theta_3\cos\Theta_1 & 0 \\ 0 & \cos\Theta_1 & 1 \end{bmatrix}.$$

Note that the introduction of the Euler angles may lead to polar singularities. For more details on *Euler angles* see [11, pp. 148-150], [70, pp. 493-495], [44, pp. 150-154], [59] and [69].

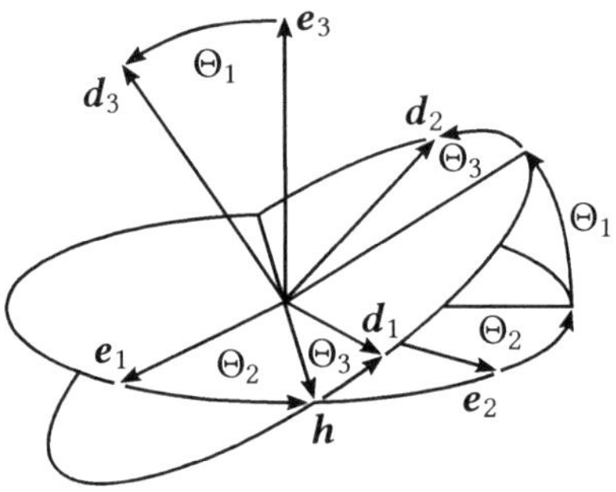

Figure 2.9.: Rotation of a moving frame $\{\boldsymbol{d}_i\}$ relative to a fixed frame $\{\boldsymbol{e}_i\}$ by using the *Euler angles*.

Focusing on hyperelastic materials, the constitutive relations are governed by the stored energy function $\Psi = \hat{\Psi}(\boldsymbol{\gamma}, \boldsymbol{\kappa})$. We assume that

$$\begin{aligned} N_i &= \partial_{\gamma_i}\Psi(\boldsymbol{\gamma}, \boldsymbol{\kappa}) = \hat{N}_i(\boldsymbol{\gamma}, \boldsymbol{\kappa}) = F_{ik}(\boldsymbol{\gamma}, \boldsymbol{\kappa})\gamma_k\,, \\ M_i &= \partial_{\kappa_i}\Psi(\boldsymbol{\gamma}, \boldsymbol{\kappa}) = \hat{M}_i(\boldsymbol{\gamma}, \boldsymbol{\kappa}) = G_{ik}(\boldsymbol{\gamma}, \boldsymbol{\kappa})\kappa_k \end{aligned} \tag{2.56}$$

holds. Note that the fundamental conditions regarding the limiting deformation cases have to be fulfilled. Hence, for $\gamma_\alpha \rightarrow \{\pm\infty\}$ the contact force N_α should tend to $\pm\infty$ and the contact force N_3 should tend to $\pm\infty$ for $\gamma_3 \rightarrow \{\infty, 0\}$. Since the *principle of impenetrability of matter* must not be violated, the contact moments M_i should tend to $\pm\infty$ as the curvature κ_i tends to an upper or lower bound, where an intersection of adjacent cross-sections is imminent (cf. Fig. 2.10).

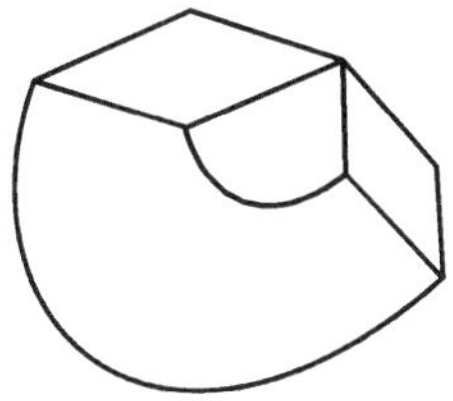

Figure 2.10.: Limiting case defined by the *principle of impenetrability of matter*: Imminent intersection of adjacent cross-sections.

Example 2.10 (Planar constitutive model). *Considering planar motion, the following strain energy function of the form*

$$\Psi = \frac{1}{2}\left(EI\kappa_2^2 + GA\gamma_1^2 + \frac{EA}{2}(\gamma_3^2 - 2\ln\gamma_3 - 1)\right) \tag{2.57}$$

might be chosen, leading to the constitutive relations for the contact forces N_1 and N_3 as well as for the contact moment M_2 as given by

$$N_3 = \frac{EA}{2}(\gamma_3 - \gamma_3^{-1}), \quad N_1 = GA\,\gamma_1 \quad \text{and} \quad M_2 = EI\,\kappa_2\,. \tag{2.58}$$

Note that the following fundamental conditions need to be ensured

$$\begin{aligned} &\hat{N}_3(\gamma_3) \longrightarrow \pm\infty \quad \text{for} \quad \gamma_3 \longrightarrow \begin{Bmatrix} \infty \\ 0 \end{Bmatrix} \\ &\hat{N}_1(\gamma_1) \longrightarrow \pm\infty \quad \text{for} \quad \gamma_1 \longrightarrow \pm\infty \\ &\hat{M}_2(\kappa_2) \longrightarrow \pm\infty \quad \text{for} \quad \kappa_2 \longrightarrow \begin{Bmatrix} sup \\ inf \end{Bmatrix} \{\kappa_2 : \gamma_3 > V(\kappa_2, s)\}\,, \end{aligned} \tag{2.59}$$

where condition (2.59)$_3$ *ensures a non-intersection of neighboring cross-sections (cf. [7]).*

By taking the constitutive relation (2.56) into account and recalling (2.39), the contact force and torque might be reformulated as

$$\boldsymbol{n} = \boldsymbol{\Lambda} \cdot F_{ik}(\boldsymbol{e}_k \otimes \boldsymbol{e}_i)\boldsymbol{\Lambda}^T \cdot \partial_s \boldsymbol{r} \quad = (\boldsymbol{\Lambda}\boldsymbol{F}^T\boldsymbol{\Lambda}^T) \cdot \partial_s \boldsymbol{r} \tag{2.60}$$

and

$$\boldsymbol{m} = \boldsymbol{\Lambda} G_{ik}(\boldsymbol{e}_k \otimes \boldsymbol{e}_i) \cdot \boldsymbol{R}^T \partial_s \boldsymbol{\Theta} \quad = (\boldsymbol{\Lambda}\boldsymbol{G}^T\boldsymbol{R}^T) \cdot \partial_s \boldsymbol{\Theta}\,, \tag{2.61}$$

respectively. In Eq. (2.60) and Eq. (2.61), $\boldsymbol{F} = F_{ik}(\boldsymbol{\gamma})(\boldsymbol{e}_i \otimes \boldsymbol{e}_k)$ $\boldsymbol{G} = G_{ik}(\boldsymbol{\gamma})(\boldsymbol{e}_i \otimes \boldsymbol{e}_k)$ has been introduced and use of the identity $\boldsymbol{Ab} \otimes \boldsymbol{Cd} = \left[\boldsymbol{C}(\boldsymbol{d} \otimes \boldsymbol{b})\boldsymbol{A}^T\right]^T$ has been made[4]. Recalling that $\partial_t \boldsymbol{d}_\alpha = \boldsymbol{\omega} \times \boldsymbol{d}_\alpha = \boldsymbol{S}_\omega \boldsymbol{d}_\alpha$, the second time derivative of the directors $\boldsymbol{d}_\alpha$ can be established as

$$\partial_t^2 \boldsymbol{d}_\alpha = \partial_t(\boldsymbol{S}_\omega \boldsymbol{d}_\alpha) = \partial_t(\boldsymbol{S}_\omega)\boldsymbol{d}_\alpha + \boldsymbol{S}_\omega^2 \boldsymbol{d}_\alpha = (\partial_t \boldsymbol{\omega} \times \boldsymbol{d}_\alpha) + \boldsymbol{S}_\omega^2 \boldsymbol{d}_\alpha = \boldsymbol{S}_\omega^2 \boldsymbol{d}_\alpha - \boldsymbol{d}_\alpha \times \partial_t \boldsymbol{\omega}\,. \tag{2.62}$$

[4] This property follows directly from the definition of the dyadic product $(\boldsymbol{a} \cdot \boldsymbol{b})\boldsymbol{c} = (\boldsymbol{c} \otimes \boldsymbol{a}) \cdot \boldsymbol{b}$ (cf. [50, p.10])

This leads to the time derivative of the linear momentum relative to $\boldsymbol{r}(s,t)$ as given by

$$\begin{aligned} \partial_t^2 \boldsymbol{q}(s,t) &= (\rho S_\alpha)(s)\left(-\boldsymbol{d}_\alpha \times \partial_t \boldsymbol{\omega} + \boldsymbol{S}_\omega^2 \boldsymbol{d}_\alpha\right) \\ &= \boldsymbol{S}_\omega^2 \boldsymbol{q} - (\boldsymbol{q} \times \partial_t \boldsymbol{\omega}) \\ &= \boldsymbol{S}_\omega^2 \boldsymbol{q} - \boldsymbol{S}_q \partial_t (\boldsymbol{R} \partial_t \Theta) \ . \end{aligned} \tag{2.63}$$

Note that $\boldsymbol{S}_\omega$ and $\boldsymbol{S}_q$, introduced in (2.62) and (2.63), respectively, denotes the skew-symmetric matrix

$$\boldsymbol{S}_{(\cdot)} = \begin{bmatrix} 0 & -(\cdot)_3 & (\cdot)_2 \\ (\cdot)_3 & 0 & -(\cdot)_1 \\ -(\cdot)_2 & (\cdot)_1 & 0 \end{bmatrix} .$$

Taking advantage of this relations, the balance of linear momentum, Eq. $(2.51)_1$ can be reformulated as

$$(\rho A)(s)\partial_t^2 \boldsymbol{r} - \boldsymbol{S}_q \boldsymbol{R} \partial_t^2 \Theta - \partial_s \left(\boldsymbol{\Lambda} \boldsymbol{F}^T \boldsymbol{\Lambda}^T \partial_s \boldsymbol{r}\right) = \bar{\boldsymbol{n}} - \boldsymbol{S}_\omega^2 \boldsymbol{q} + \boldsymbol{S}_q \partial_t \boldsymbol{R} \partial_t \Theta \ . \tag{2.64}$$

Furthermore, by consulting (2.49) for a definition of the angular momentum relative to $\boldsymbol{r}(s,t)$, its time derivative might be rewritten as

$$\begin{aligned} \partial_t \boldsymbol{h}(s,t) &= (\rho I_{\alpha\beta})(s)\partial_t(\boldsymbol{d}_\alpha \times \partial_t \boldsymbol{d}_\beta) \\ &= (\rho I_{\alpha\beta})(s)\left[\boldsymbol{d}_\alpha \times (\partial_t \boldsymbol{\omega} \times \boldsymbol{d}_\beta) + \boldsymbol{d}_\alpha \times (\boldsymbol{\omega} \times (\boldsymbol{\omega} \times \boldsymbol{d}_\beta)) + (\boldsymbol{\omega} \times \boldsymbol{d}_\alpha) \times (\boldsymbol{\omega} \times \boldsymbol{d}_\beta)\right] \\ &= (\rho I_{\alpha\beta})(s)[\boldsymbol{d}_\alpha \times \left(\partial_t(\boldsymbol{R}\partial_t \Theta) \times \boldsymbol{d}_\beta\right) + \boldsymbol{d}_\alpha \times (\boldsymbol{\omega} \times (\boldsymbol{\omega} \times \boldsymbol{d}_\beta)) \\ &\qquad + (\boldsymbol{\omega} \times \boldsymbol{d}_\alpha) \times (\boldsymbol{\omega} \times \boldsymbol{d}_\beta)] \ , \end{aligned} \tag{2.65}$$

such that the balance of angular momentum (2.52) can be reformulated as

$$\boldsymbol{S}_q \partial_t^2 \boldsymbol{r} - (\rho \boldsymbol{J})(s)\partial_t^2 \Theta - \partial_s \left(\boldsymbol{\Lambda} \boldsymbol{G}^T \boldsymbol{R}^T \cdot \partial_s \Theta\right) = \partial_s \boldsymbol{r} \times \boldsymbol{n} + \bar{\boldsymbol{m}} - (\rho \boldsymbol{k})(s) \ , \tag{2.66}$$

where the functions $\boldsymbol{J} : \Omega \mapsto \mathbb{R}^{3,3}$ and $\boldsymbol{k} : \Omega \mapsto \mathbb{R}^3$ have been introduced as[5]

$$\boldsymbol{J}(s) \equiv (I_{\alpha\beta})(s)\boldsymbol{S}_{\boldsymbol{d}_\alpha}\boldsymbol{S}_{\boldsymbol{d}_\beta} \tag{2.67}$$

and

$$\boldsymbol{k}(s) \equiv (I_{\alpha\beta})(s)\left(\boldsymbol{d}_\alpha \times \boldsymbol{S}_\omega^2 \boldsymbol{d}_\beta + \boldsymbol{\omega} \cdot (\boldsymbol{d}_\alpha \times \boldsymbol{d}_\beta)\boldsymbol{\omega}) + \boldsymbol{d}_\alpha \times \left(\partial_t \boldsymbol{R} \partial_t \Theta \times \boldsymbol{d}_\beta\right)\right) \ . \tag{2.68}$$

In the context of the inverse dynamics problem, the actuating force and torque

$$\boldsymbol{f}, \boldsymbol{m} : \partial\Omega_f \mapsto \mathbb{R}^d \tag{2.69}$$

applied to $\partial\Omega_f = \{0\} \times \mathcal{T}$, causing a partly specified motion

$$\boldsymbol{x}|_{(s,t)\in\partial\Omega_\eta} \equiv \boldsymbol{\gamma}(t) : \partial\Omega_\eta \mapsto \mathbb{R}^d. \tag{2.70}$$

[5] Note, that use of the $(\boldsymbol{a} \times \boldsymbol{b}) \times (\boldsymbol{a} \times \boldsymbol{c}) = \boldsymbol{a} \cdot (\boldsymbol{b} \times \boldsymbol{c})\boldsymbol{a}$ has been made.

on $\partial\Omega_\eta = \{L\} \times \mathcal{T}$ are searched for. (cf. Fig. 2.11 for a planar illustration of the inverse dynamics problem)

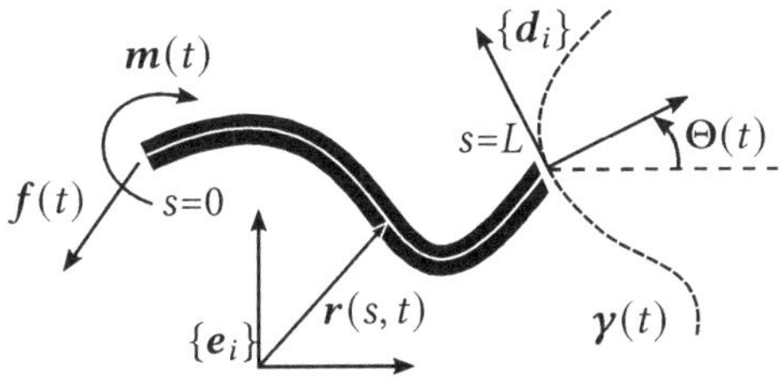

Figure 2.11.: Planar illustration of the inverse dynamics problem: Find the force $\boldsymbol{f}(t)$ and moment $\boldsymbol{m}(t)$ acting at s=0 such that the imposed time-varying constraint on $\boldsymbol{d}_i(t)$ and $\boldsymbol{\Theta}(t)$ at $s = L$ is met.

Regarding multibody systems, *Neumann* boundary conditions in the form of a system of ordinary differential equations

$$\boldsymbol{B}\partial_s\boldsymbol{x}|_{(s,t)\in\partial\Omega_\eta} = \boldsymbol{\eta}(\partial_t^2\boldsymbol{x}, \partial_t\boldsymbol{x}, \boldsymbol{x}, t) \tag{2.71}$$

may be taken into account on $\partial\Omega_\gamma = \{L\} \times \mathcal{T}$. By specifying initial conditions on $\partial\Omega_0 = S \times \{0\}$

$$\boldsymbol{x}|_{(s,t)\in\partial\Omega_0} = \boldsymbol{x}_0, \qquad \partial_t\boldsymbol{x}|_{(s,t)\in\partial\Omega_0} = \boldsymbol{v}_0 \tag{2.72}$$

and introducing the coefficients

$$A = \begin{bmatrix} \rho A I & -S_q \\ S_q & -\rho J \end{bmatrix}, \quad B = \begin{bmatrix} \boldsymbol{\Lambda F}^T\boldsymbol{\Lambda}^T & \boldsymbol{0} \\ \boldsymbol{0} & \boldsymbol{\Lambda G}^T\boldsymbol{R}^T \end{bmatrix} \quad \text{and} \quad C = \begin{bmatrix} \bar{\boldsymbol{n}} - S_\omega^2\boldsymbol{q} + S_q\partial_t R\partial_t\Theta \\ (\partial_s\boldsymbol{r} \times \boldsymbol{n}) + \bar{\boldsymbol{m}} - \rho\boldsymbol{k} \end{bmatrix} \tag{2.73}$$

the problem aligns with the general formulation of the inverse dynamics of flexible mechanical systems proposed in Sec. 1.1.

Example 2.11 (Planar problem). *Subsequently, the special case of planar motion by additionally assuming straight reference configurations as well as $\boldsymbol{r} = \boldsymbol{r}_S$ is considered. Then strain energy functions of the form* (2.57) *can be assumed (cf. Ex. 2.10) such that the functions $\boldsymbol{F}(\boldsymbol{\gamma})$ and $\boldsymbol{G}(\boldsymbol{\kappa})$, introduced in* (2.60) *and* (2.61)*, respectively, appear in diagonal form, i.e. $\boldsymbol{F} = \mathbf{diag}\left(GA, 0, \frac{EA}{2}\left(1 - \gamma_3^{-2}\right)\right)$ and $\boldsymbol{G} = \mathbf{diag}(0, EI, 0)$. Consequently, the diagonal block matrices $\boldsymbol{\Lambda F}^T\boldsymbol{\Lambda}^T$ and $\boldsymbol{\Lambda G}^T\boldsymbol{R}^T$ in* (2.73)$_2$ *are obtained as*

$$\boldsymbol{\Lambda F}^T\boldsymbol{\Lambda}^T = GA\begin{bmatrix} \cos^2\Theta & 0 & \cos\Theta\sin\Theta \\ 0 & 0 & 0 \\ \cos\Theta\sin\Theta & 0 & \sin^2\Theta \end{bmatrix} + \frac{EA}{2}\left(1 - \gamma_3^{-2}\right)\begin{bmatrix} \sin^2\Theta & 0 & -\cos\Theta\sin\Theta \\ 0 & 0 & 0 \\ -\cos\Theta\sin\Theta & 0 & \cos^2\Theta \end{bmatrix}$$

and

$$\boldsymbol{\Lambda G}^T \boldsymbol{R} = \begin{bmatrix} 0 & 0 & 0 \\ 0 & EI & 0 \\ 0 & 0 & 0 \end{bmatrix} .$$

Note that the Euler angles Θ_2 and Θ_3 vanishes in the plane case. $\boldsymbol{\Lambda}$ has been taken from Ex. 2.8 such that the coefficient $\boldsymbol{B}$ in (2.73) *can be identified as*

$$\boldsymbol{B} = \frac{EA}{2}(1 - \nu^{-2}) \begin{bmatrix} \cos^2\Theta & \cos\Theta\sin\Theta & 0 \\ \cos\Theta\sin\Theta & \sin^2\Theta & 0 \\ 0 & 0 & 0 \end{bmatrix} + \begin{bmatrix} GA\sin^2\Theta & -GA\cos\Theta\sin 2\Theta & 0 \\ -GA\cos\Theta\sin 2\Theta & GA\cos^2\Theta & 0 \\ 0 & 0 & EI \end{bmatrix} .$$

Since the functions $\boldsymbol{J}$ and $\boldsymbol{k}$, introduced in (2.67) *and* (2.68)*, respectively, reduce to*

$$\boldsymbol{J} = \rho I_{11} \boldsymbol{S}^2_{\boldsymbol{d}_1} = \rho I_{11} \begin{bmatrix} 0 & 0 & \cdot \\ \cdot & -(d_{11}^2 + d_{13}^2) = -1 & \cdot \\ \cdot & 0 & 0 \end{bmatrix} \quad \text{and} \quad \boldsymbol{k} = \boldsymbol{0},$$

the two remaining coefficients $\boldsymbol{A}$ and $\boldsymbol{C}$ in (2.73) *can be obtained as*

$$\boldsymbol{A} = \begin{bmatrix} \rho A & 0 & 0 \\ 0 & \rho A & 0 \\ 0 & 0 & \rho I \end{bmatrix} \quad \text{and} \quad \boldsymbol{C} = \begin{bmatrix} \bar{n}_1 \\ \bar{n}_3 \\ \bar{m}_2 + (\partial_s r_3 n_1 - \partial_s r_1 n_3) \end{bmatrix} \tag{2.74}$$

Therein, use of

$$\partial_s \boldsymbol{r} \times \boldsymbol{n} = \begin{bmatrix} \partial_s r_1 \\ 0 \\ \partial_s r_3 \end{bmatrix} \times \begin{bmatrix} n_1 \\ 0 \\ n_3 \end{bmatrix} = \begin{bmatrix} 0 \\ \partial_s r_3 n_1 - \partial_s r_1 n_3 \\ 0 \end{bmatrix}$$

has been made.

Remark 2.12. *(Componential representation) The classical equations of geometrically exact beams have been introduced in* $(2.51)_1$ *and* (2.52) *in vectorial representation. However, in certain cases, a componential representation of these equations might be more convenient*[6].

The contact forces and torques might then be established as

$$\boldsymbol{n} = N_i \boldsymbol{d}_i \quad \text{and} \quad \boldsymbol{m} = M_i \boldsymbol{d}_i \ , \tag{2.75}$$

respectively. By differentiating Eq. $(2.75)_1$ *with respect to the spatial variable s*

$$\partial_s \boldsymbol{n} = (\partial_s N_i)\boldsymbol{d}_i + N_i \partial_s \boldsymbol{d}_i = (\partial_s N_i)\boldsymbol{d}_i + N_i(\boldsymbol{\kappa} \times \boldsymbol{d}_i) \ ,$$

[6] The terminology *vectorial* and *componential* representation for distinguishing the chosen basis can be found in [7].

the balance of linear momentum can be stated as

$$\partial_s N_i \delta_{ij} + N_i(\boldsymbol{\kappa} \times \boldsymbol{d}_i) \cdot \boldsymbol{d}_j + \bar{\boldsymbol{n}} \cdot \boldsymbol{d}_i = \partial_t \boldsymbol{p} \cdot \boldsymbol{d}_i \ . \tag{2.76}$$

Taking advantage of the property of the triple product $(\boldsymbol{\kappa} \times \boldsymbol{d}_i) \cdot \boldsymbol{d}_j = \boldsymbol{\kappa} \cdot (\boldsymbol{d}_i \times \boldsymbol{d}_j)$, *Eq.* (2.76) *might be reshaped as*

$$\partial_s N_i \delta_{ij} + N_i Q_{ij} + \bar{\boldsymbol{n}} \cdot \boldsymbol{d}_j = \partial_t \boldsymbol{p} \cdot \boldsymbol{d}_j \tag{2.77}$$

by introducing

$$Q_{ij} \equiv \boldsymbol{\kappa} \cdot (\boldsymbol{d}_i \times \boldsymbol{d}_j) \ .$$

Differentiating Eq. $(2.75)_2$ *with respect to the spatial variable s*

$$\partial_s \boldsymbol{m} = (\partial_s M_i)\boldsymbol{d}_i + M_i(\boldsymbol{\kappa} \times \boldsymbol{d}_i)$$

and taking into account, that[7]

$$\partial_s \boldsymbol{r} \times \boldsymbol{n} = (S^{\gamma} \boldsymbol{N})\boldsymbol{d}_i$$

holds, the balance of angular momentum might be established in componential form as

$$\partial_s M_i \delta_{ij} + M_i Q_{ij} + N_i S^{\gamma}_{ji} + \bar{\boldsymbol{m}} \cdot \boldsymbol{d}_j = \partial_t \hat{\boldsymbol{l}} \cdot \boldsymbol{d}_j . \tag{2.78}$$

Considering that $\bar{\boldsymbol{n}} \cdot \boldsymbol{d}_i = \bar{\boldsymbol{n}}(\boldsymbol{\Lambda}^T \boldsymbol{e}_i) = (\boldsymbol{\Lambda}^T \bar{\boldsymbol{n}})\boldsymbol{e}_i$, *the equations* (2.77) *and* (2.78) *constitute the equations of motion of geometrically exact beams in componential form, as given by*

$$\begin{bmatrix} N_i \\ M_i \end{bmatrix}_{,s} + \begin{bmatrix} Q_{ij} & 0 \\ S^{\gamma}_{ji} & Q_{ij} \end{bmatrix} \begin{bmatrix} N_i \\ M_i \end{bmatrix} + \begin{bmatrix} \bar{\boldsymbol{n}} \cdot \boldsymbol{d}_j \\ \bar{\boldsymbol{m}} \cdot \boldsymbol{d}_j \end{bmatrix} = \begin{bmatrix} \partial_t \boldsymbol{p} \cdot \boldsymbol{d}_j \\ \partial_t \hat{\boldsymbol{l}} \cdot \boldsymbol{d}_j \end{bmatrix} . \tag{2.79}$$

Example 2.13 (Componential representation). *Regarding planar motion, the functions* $\boldsymbol{Q}$ *and* $\boldsymbol{S}^{\gamma}$, *introduced in* (2.77) *and* (2.78), *reduce to*

$$\boldsymbol{Q} = \begin{bmatrix} 0 & 0 & \kappa_2 \\ 0 & 0 & 0 \\ -\kappa_2 & 0 & 0 \end{bmatrix} \qquad \text{and} \qquad S^{\gamma} = \begin{bmatrix} 0 & -\gamma_3 & 0 \\ \gamma_3 & 0 & -\gamma_1 \\ 0 & \gamma_1 & 0 \end{bmatrix}$$

and the equations of motion in componential form (2.79), *simplify to (cf. [7, Chap. IV.1.])*

$$\begin{bmatrix} N_1 \\ N_3 \\ M_2 \end{bmatrix}_{,s} + \begin{bmatrix} 0 & \kappa_2 & 0 \\ -\kappa_2 & 0 & 0 \\ \gamma_3 & -\gamma_1 & 0 \end{bmatrix} \begin{bmatrix} N_1 \\ N_3 \\ M_2 \end{bmatrix} + \begin{bmatrix} \bar{\boldsymbol{n}} \cdot \boldsymbol{d}_1 \\ \bar{\boldsymbol{n}} \cdot \boldsymbol{d}_3 \\ \bar{\boldsymbol{m}} \cdot \boldsymbol{d}_2 \end{bmatrix} = \begin{bmatrix} \partial_t \boldsymbol{p} \cdot \boldsymbol{d}_1 \\ \partial_t \boldsymbol{p} \cdot \boldsymbol{d}_3 \\ \partial_t \hat{\boldsymbol{l}} \cdot \boldsymbol{d}_2 \end{bmatrix} . \tag{2.80}$$

Using the hyperelastic constitutive equation, proposed in Ex. 2.10, a relation for the forces and torques to the strain variables can be established as

$$N_i = F_{ij}\gamma_j = F_{ij}(\partial_s \boldsymbol{r} \cdot \bar{\boldsymbol{\Lambda}} \boldsymbol{e}_j) = \begin{bmatrix} GA\cos\Theta & GA\sin\Theta \\ -\frac{EA}{2}(1-\gamma_3^{-2})\sin\Theta & \frac{EA}{2}(1-\gamma_3^{-2})\cos\Theta \end{bmatrix} \quad \text{for} \quad i \in \{1,3\}$$

[7] This can be shown by simply computing $\partial_s \boldsymbol{r} \times \boldsymbol{n} = \gamma_k \boldsymbol{d}_k \times N_i \boldsymbol{d}_i = (\gamma_1 \boldsymbol{d}_1 + \cdots) \times (N_1 \boldsymbol{d}_1 + \cdots) = (\boldsymbol{\gamma} \times \boldsymbol{N})\boldsymbol{d}_j$

and

$$M_i = G_{ij}\kappa_j = EI\partial_s\Theta \quad \textit{for} \quad i = 2 \; .$$

This leads to the characterizing coefficients

$$\boldsymbol{A} = \rho A \begin{bmatrix} \cos\Theta & \sin\Theta & 0 \\ -\sin\Theta & \cos\Theta & 0 \\ 0 & 0 & \rho I_{11} \end{bmatrix} \quad \textit{and} \quad \boldsymbol{C} = \begin{bmatrix} 0 & \kappa_2 & 0 \\ -\kappa_2 & 0 & 0 \\ \gamma_3 & -\gamma_1 & 0 \end{bmatrix} \begin{bmatrix} N_1 \\ N_3 \\ M_2 \end{bmatrix} + \begin{bmatrix} \bar{\boldsymbol{n}} \cdot \boldsymbol{d}_1 \\ \bar{\boldsymbol{n}} \cdot \boldsymbol{d}_1 \\ \bar{\boldsymbol{m}} \cdot \boldsymbol{d}_2 \end{bmatrix}$$

as well as

$$\boldsymbol{B} = \begin{bmatrix} \frac{EA}{2}(1 - \nu^{-2})\cos\Theta & \frac{EA}{2}(1 - \nu^{-2})\sin\Theta & 0 \\ -GA\sin\Theta & GA\cos\Theta & 0 \\ 0 & 0 & EI \end{bmatrix} \; .$$

Hence, the problem at hand again coincides with the quasi-linear hyperbolic partial differential equation, proposed in Sec. 1.1.

Remark 2.14 (From Simo-Reissner via Timoshenko-Ehrenfest to Bernoulli-Euler). *Introducing the function $\boldsymbol{u} : \Omega \mapsto \mathbb{R}$ denoting the displacement relative to the reference configuration $\boldsymbol{R}(s)$, the spatial configuration of the beam can be obtained as*

$$\boldsymbol{r}(s,t) = \boldsymbol{R}(s) + \boldsymbol{u}(s,t) \; . \tag{2.81}$$

Assuming straight reference configurations, i.e. $\boldsymbol{R}(s) = s\ \boldsymbol{e}_3$, the derivative of (2.81) *with respect to the spatial variable s can be determined to be $\partial_s\boldsymbol{r} = \boldsymbol{e}_3 + \partial_s\boldsymbol{u}$. Then the strain variables γ_1 and γ_1 can be established as*

$$\begin{aligned} \gamma_3 = \partial_s\boldsymbol{r}\cdot\boldsymbol{d}_3 &= (\partial_s\boldsymbol{u} + \boldsymbol{e}_3)\cdot\boldsymbol{d}_3 = (\partial_s\boldsymbol{u} + \boldsymbol{e}_3)\cdot(\cos\Theta\boldsymbol{e}_3 + \sin\Theta\boldsymbol{e}_1) \\ &= (\partial_s u_3 + 1)\cos\Theta + \partial_s u_1 \sin\Theta = \Gamma_3 + 1 \end{aligned} \tag{2.82}$$

and

$$\begin{aligned} \gamma_1 = \partial_s\boldsymbol{r}\cdot\boldsymbol{d}_1 &= (\partial_s\boldsymbol{u} + \boldsymbol{e}_3)\cdot\boldsymbol{d}_1 = (\partial_s\boldsymbol{u} + \boldsymbol{e}_3)\cdot(-\sin\Theta\boldsymbol{e}_3 + \cos\Theta\boldsymbol{e}_1) \\ &= -(\partial_s u_3 + 1)\sin\Theta + \partial_s u_1 \cos\Theta = \Gamma_1 \; , \end{aligned} \tag{2.83}$$

respectively (cf. [88, Eq. (5.2)]). Assuming inextensibility of the beam, i.e.

$$\gamma_3 = \partial_s u_3 \cos\Theta + \cos\Theta + \partial_s u_1 \sin\Theta \equiv 1 \; , \tag{2.84}$$

implies a relation for the spatial derivative of the displacement u_3, given by

$$\partial_s u_3 = \frac{1}{\cos\Theta}(1 - \cos\Theta - \partial_s u_1 \sin\Theta) \; . \tag{2.85}$$

Inserting (2.85) *into* (2.83)*, the strain variable γ_1 is obtained as*

$$\gamma_1 = -\frac{\sin\Theta}{\cos\Theta} + \partial_s u_1 \left(\frac{\sin^2\Theta}{\cos\Theta} + \cos\Theta\right) = \frac{1}{\cos\Theta}(\partial_s u_1 - \sin\Theta) \; . \tag{2.86}$$

Inextensibility of the beam implies that the force N_3 is not defined constitutively anymore. Instead N_3 has to be considered as a Lagrange multiplier enforcing condition (2.84). *For small deformations it is feasible to assume* $N_3 = 0$. *Then Eq.* (2.80) *immediately reduces to*

$$\partial_s N_1 + \bar{n}_1 = \partial_t p_1 \quad \text{and} \quad \partial_s M_2 + N_1 + \bar{m}_2 = \partial_t \hat{l}_2 \; . \tag{2.87}$$

The system of partial differential equations (2.87) *can be traced back to the pioneering work of S.P. Timoshenkho and P. Ehrenfest [95, 96].*

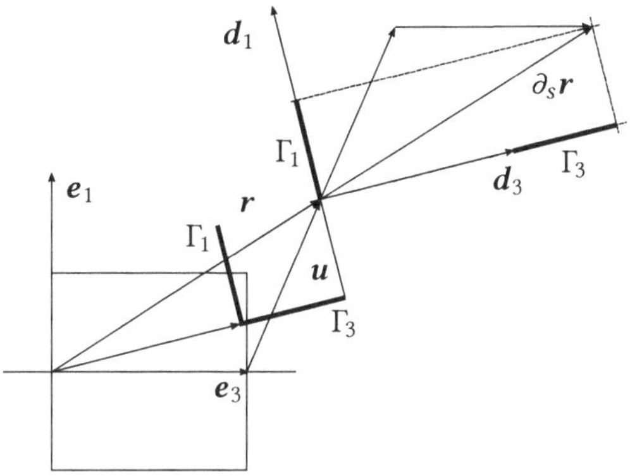

Figure 2.12.: Illustration of the strain components of the geometrically exact beam formulation and its physical interpretation (cf. [91]).

A further restriction, pertaining to the rigidity of shear deformation, i.e. $\gamma_1 \equiv 0$ *and simultaneously neglecting rotatory inertia effects, leads to the classical equation of motion*

$$\partial_s(\partial_s M_2 + \bar{m}_2) + \bar{n}_1 = \partial_t p_1 \; . \tag{2.88}$$

Eq. (2.88) *has been developed originally by L. Euler, Jas. Bernoulli and D. Bernoulli, by additionally assuming linear constitutive relations of the form*

$$M = EI\kappa = EI\partial_s\Theta = EI\partial_s^2 u_1 \; . \tag{2.89}$$

2.3. General nonlinear continuum formulation

After having considered slender structures such as strings and beams in Sec. 2.1 and Sec. 2.2, respectively, the elastodynamics of general *three*-dimensional continua will be considered in this section. Therefore, we aim to elaborate briefly the cornerstones of the underlying theory including large deformations. Analogously to the theory of slender structures, it might be convenient to start with a recapitulation of the classical concepts of the kinematics of continua, before fundamental balance equations, governing the motion of arbitrary deformable bodies $\mathcal{B}$, can be introduced.

Kinematics. General α-dimensional continua might be perceived as arbitrary shaped and deformable bodies occupying regions in α-dimensional space, i.e. $\mathcal{B} \subset \mathbb{R}^\alpha$. We will focus on formulations based on *material* points $s \in \mathcal{B}$ at time $t \in \mathcal{T}$, i.e. material coordinates $(s,t) \in \Omega$ are chosen as independent variables. Such formulations are frequently referred to as *material* formulations, or more historically *Lagrangian* formulations[8].

The position of every material point $s \in \mathcal{B}$ for every time $t \in \mathcal{T}$ is then defined by the injective[9] map

$$(s,t) \in \Omega \mapsto \boldsymbol{r}(s,t) \in \mathbb{R}^\alpha \, . \tag{2.90}$$

Similar to the *one*-dimensional case of slender structures such as strings and beams, it is inevitable to find feasible measurements in order to quantify the deformation of the body. Therefore, we will investigate the change of a curve $s_0(\xi) : \xi \in [a,b] \mapsto \mathbb{R}^\alpha$, parameterized by the arc-length parameter ξ (see Fig. 2.13).

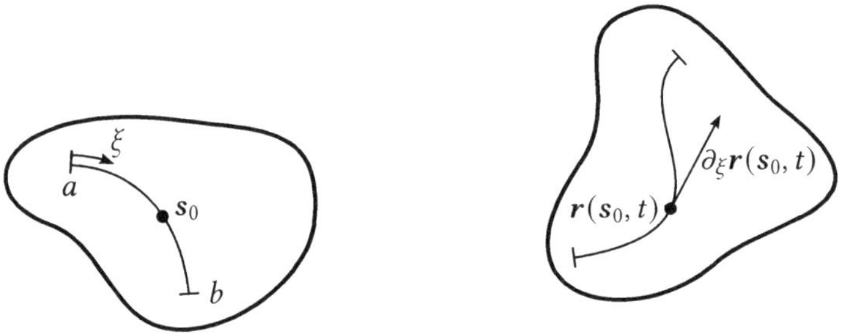

Figure 2.13.: Kinematics of a general non-linear continuum formulation.

By defining the *deformation gradient* $\boldsymbol{F} \equiv \partial_s \boldsymbol{r}(s_0,t)$, the vector field tangential to the curve s_0 is obtained as

$$\partial_\xi \boldsymbol{r}(s_0,t) = \partial_s \boldsymbol{r}(s_0(\xi),t) \cdot \mathrm{D}_\xi s_0 = \boldsymbol{F}(s_0,t) \cdot \mathrm{D}_\xi s_0 \, .$$

In order to describe the change of the shape of the curve s locally, we can compare the length of the curve s_0 after some deformation

$$l = \int_a^b \left\| \partial_\xi \boldsymbol{r}(s_0,t) \right\| \, \mathrm{d}\xi = \int_a^b \left\| \partial_s \boldsymbol{r}(s_0,t) \cdot \mathrm{D}_\xi s \right\| \, \mathrm{d}\xi = \int_a^b \left\| \boldsymbol{F}(s_0,t) \cdot \mathrm{D}_\xi s \right\| \, \mathrm{d}\xi \tag{2.91}$$

to its length in reference configuration

$$l_0 = \int_a^b \left\| \mathrm{D}_\xi s_0 \right\| \, \mathrm{d}\xi$$

[8] Alternatively, the problem can be formulated in terms of the positions $\boldsymbol{r}(\boldsymbol{q}(\boldsymbol{y},t),t)$ of material points covering positions $\boldsymbol{y}$ at time t, i.e. spatial coordinates $(\boldsymbol{y},t)$ are chosen as independent variables. Therefore, this formulation is frequently referred to as *spatial* or historically *Eulerian* formulation.

[9] Injectivity of $\boldsymbol{r}(s,t)$ needs to be supposed, due to the *principle of impenetrability of matter*, i.e. two distinct material points cannot simultaneously occupy the same position in space.

and analyze the limit case of its ratio, i.e.

$$\begin{aligned}\lim_{b\to a}\frac{l}{l_0} = \left\|F(s_0,t)\cdot\partial_\xi s\right\| &= \left[(F\cdot \mathrm{D}_\xi s_0)\cdot(F\cdot \mathrm{D}_\xi s_0)\right]^{\frac{1}{2}} \\ &= \left[\mathrm{D}_\xi s_0\cdot F^T F\cdot \mathrm{D}_\xi s_0\right]^{\frac{1}{2}} \\ &= \left[\mathrm{D}_\xi s(a,t)\cdot C\cdot \mathrm{D}_\xi s(a)\right]^{\frac{1}{2}}\ .\end{aligned} \tag{2.92}$$

Thus, a fiber in point a, that is oriented tangentially to the curve s changes its length relative to its reference length according to a scalar parameter

$$\nu = \left[\mathrm{D}_\xi s(a,t)\cdot C\cdot \mathrm{D}_\xi s(a)\right]^{\frac{1}{2}}\ .$$

The curve s is stretched or compressed locally in point a as $\nu > 1$ and $\nu < 1$, respectively. For $\nu = 1$, the length of the curve does not change (cf. Sec. 2.1, for conspicuous analogy). In (2.92) the symmetric and positive definite tensor $C \equiv F^T F$ has been introduced. This tensor is frequently referred to as right *Cauchy-Green* deformation tensor.

Based on the right *Cauchy-Green* deformation tensor C, a vast variety of different strain measurements can be established. One prominent member of this family of strain measurements is the *material strain tensor*

$$E = \frac{1}{2}(C - I)$$

also referred to as *Green-Lagrange* strain tensor. Its virtue lies in the fact that it vanishes in the reference configuration.

Equilibrium. In analogy to Sec. 2.1 and Sec. 2.2, the two fundamental balance equations regarding linear and angular momentum are introduced subsequently for the general continuum formulation.

Balance of linear momentum. According to Newton's second axiom, the resultant force $f : \Omega \mapsto \mathbb{R}^\alpha$ as given by

$$f = \int_{\mathcal{B}} \mathrm{d}f(s,t)$$

acting on a body $\mathcal{B}$ with total mass

$$m(\mathcal{B}) = \int_{\mathcal{B}} \mathrm{d}m(s,t)$$

is equal to the temporal change of the linear momentum of the body as given by

$$\frac{\mathrm{d}}{\mathrm{d}t}\int_{\mathcal{B}} \partial_t r(s,t)\ \mathrm{d}m(s,t) - f(t) = 0\ . \tag{2.93}$$

Furthermore, we assume, that a decomposition of the resultant force $\boldsymbol{f}$ into body forces $\boldsymbol{b} : \Omega \mapsto \mathbb{R}^{\alpha}$ per unit of reference volume and surface traction $\boldsymbol{t} : \Omega \mapsto \mathbb{R}^{\alpha}$ per unit of reference area $\partial\mathcal{B}$, i.e.

$$\boldsymbol{f} = \int_{\mathcal{B}} \boldsymbol{b}(\boldsymbol{s}, t)\ \mathrm{d}v(\boldsymbol{s}, t) + \int_{\partial\mathcal{B}} \boldsymbol{t}(\boldsymbol{s}, t)\ \mathrm{d}a(\boldsymbol{s}, t) \tag{2.94}$$

can be applied. Eq. (2.93) along with Eq. (2.94) represent the balance equation of linear momentum in its integral form.

Local form of the balance of linear momentum. With the help of *Cauchy's theorem* the balance equations can be established in local form. This fundamental stress theorem, postulated by *Cauchy*, states that there exists locally a second-order tensor field $\boldsymbol{P} : \Omega \mapsto \mathbb{R}^{\alpha,\alpha}$ such that

$$\boldsymbol{t} = \boldsymbol{P} \cdot \boldsymbol{\nu} \tag{2.95}$$

holds. Therein, $\boldsymbol{\nu}$ denotes the unit normal vector on the surface, on which the traction $\boldsymbol{t}$ acts. To prove Eq. (2.95), the following geometrical considerations are made: Consider a tetrahedron, that is bounded by four planes with normal vectors $\boldsymbol{e}_k$ for $k \in \{1, 2, 3\}$ and $\boldsymbol{\nu}$ such that $\boldsymbol{\nu} \cdot \boldsymbol{e}_k > 0\ \forall\ k \in \{1, 2, 3\}$. The size of the tetrahedron is defined by $h\boldsymbol{\nu}$, where $h \in \mathbb{R}^{+}$. The area of the plane Π that is orthogonal to $\boldsymbol{\nu}$ can be obtained as ch^2. Accordingly the planes Π_k, that are orthogonal to $\boldsymbol{e}_k$ have the area $ch^2\boldsymbol{\nu} \cdot \boldsymbol{e}_k$. This is the projection of the area of plane Π onto plane Π_k. *Cauchy's theorem* can then basically be proven by establishing the balance of linear momentum for this specific tetrahedron as given by

$$\int_{\partial\mathcal{B}(h)} \boldsymbol{t}(\boldsymbol{s}, t, \boldsymbol{\nu})\ \mathrm{d}a(\boldsymbol{s}) + \sum_{k=1}^{3} \int_{\partial\mathcal{B}_k(h)} \boldsymbol{t}(\boldsymbol{s}, t, -\boldsymbol{e}_k)\ \mathrm{d}a(\boldsymbol{s}) = \int_{\mathcal{B}(h)} \rho\partial_t^2\boldsymbol{r}(\boldsymbol{s}, t) - \boldsymbol{f}(\boldsymbol{s}, t)\ \mathrm{d}v(\boldsymbol{s})\ . \tag{2.96}$$

Dividing (2.96) by ch^2 and evaluating the limit for $h \longrightarrow 0$ gives for every material point rise to

$$\boldsymbol{t}(\boldsymbol{s}_0, t, \nu) = -\sum_{k=1}^{3} (\boldsymbol{t}(\boldsymbol{s}_0, t, -\boldsymbol{e}_k) \otimes \boldsymbol{e}_k) \cdot \boldsymbol{\nu}$$

and *Cauchy's theorem* (2.95) is proven to be right, by introducing the second-order tensor field

$$\boldsymbol{P}(\boldsymbol{s}, t) \equiv -\sum_{k=1}^{3} (\boldsymbol{t}(\boldsymbol{s}, t, -\boldsymbol{e}_k) \otimes \boldsymbol{e}_k)\ .$$

This second-order tensor field is classically referred to as *first Piola-Kirchhoff stress tensor field.* Due to its lack of symmetry it sometimes proves convenient to introduce the *second Piola Kirchhoff stress tensor* as $\boldsymbol{S} \equiv \boldsymbol{F}\boldsymbol{P}$ or the *Cauchy stress tensor* as $\boldsymbol{\sigma} \equiv (\det \boldsymbol{F})^{-1}\boldsymbol{P}\boldsymbol{F}^{T}$.

Inserting *Cauchy's theorem* (2.95) along with (2.94) into (2.93) and applying the divergence theorem, a local form of the balance of linear momentum is given by

$$\operatorname{div} \boldsymbol{P}^{T} + \boldsymbol{b} = \rho\partial_t^2\boldsymbol{r}\ . \tag{2.97}$$

Balance of angular momentum. Besides the balance of linear momentum, the balance of angular momentum represents the second fundamental principle for elastodynamics, i.e.

$$\frac{\mathrm{d}}{\mathrm{d}t}\int_{\mathcal{B}} \boldsymbol{r}(\Omega) \times \partial_t \boldsymbol{r}(\Omega)\ \mathrm{d}m(s) - \boldsymbol{m}(t, \mathbf{0}) = \mathbf{0}\ . \tag{2.98}$$

Therein, $\boldsymbol{m}(t, \mathbf{0})$ denotes the resultant torque on the body $\mathcal{B}$ about the origin $\mathbf{0}$ at time t as given by

$$\boldsymbol{m}(t, \mathbf{0}) = \int_{\mathcal{B}} \boldsymbol{r}(\Omega) \times \mathrm{d}\boldsymbol{f}(\Omega) = \int_{\mathcal{B}} \boldsymbol{r}(\Omega) \times \boldsymbol{b}\, \mathrm{d}v(s) + \int_{\partial\mathcal{B}} \boldsymbol{t}(\Omega)\, \mathrm{d}a(s)\ . \tag{2.99}$$

Analogously to the balance of linear momentum, the balance of angular momentum can be described in its local form too. This implies

$$\boldsymbol{F} \cdot \boldsymbol{P}^T = \boldsymbol{P} \cdot \boldsymbol{F}^T\ . \tag{2.100}$$

This might be proven by applying *Cauchy's theorem* to the balance equations (2.98) and (2.99) straightforward obtaining

$$\int_{\partial\mathcal{B}} \boldsymbol{r} \times (\boldsymbol{P} \cdot \boldsymbol{\nu})\ \mathrm{d}a + \int_{\mathcal{B}} (\boldsymbol{r} \times \boldsymbol{f} - \rho \boldsymbol{r} \times \partial_t^2 \boldsymbol{r})\ \mathrm{d}v = \mathbf{0}\ . \tag{2.101}$$

Multiplying (2.101) with the cross product of two arbitrary but constant vectors $\boldsymbol{a} \in \mathbb{R}^\alpha$ and $\boldsymbol{b} \in \mathbb{R}^\alpha$, Eq. (2.101) reduces to

$$(\boldsymbol{b} \otimes \boldsymbol{a} - \boldsymbol{a} \otimes \boldsymbol{b}) : \int_{\mathcal{B}} \boldsymbol{P} \cdot \boldsymbol{F}^T + (\mathrm{div}\,(\boldsymbol{P}^T) + \boldsymbol{f} - \rho \partial_t^t \boldsymbol{r}) \otimes \boldsymbol{r}\ \mathrm{d}v = 0\ . \tag{2.102}$$

Keeping in mind that $\mathcal{B}$ is arbitrary and using the local form of the balance of linear momentum (2.97), the following relation

$$(\boldsymbol{b} \otimes \boldsymbol{a} - \boldsymbol{a} \otimes \boldsymbol{b}) : \boldsymbol{P} \cdot \boldsymbol{F}^T$$

is obtained and the symmetry postulated in (2.100) is proven to be true.

The partial differential equation (2.97) along with (2.100) constitute the material form of the classical equations of motion for general continua and aligns with the problem formulation postulated in Sec. 1.1 for

$$\boldsymbol{A} = \rho \boldsymbol{I}, \qquad \boldsymbol{B}\partial_s \boldsymbol{r} = \boldsymbol{P} \quad \text{and} \quad \boldsymbol{C} = \boldsymbol{b}$$

With (2.97) and (2.100), *six* equations could have been derived for the *twelve* components of the unknown functions $\boldsymbol{r}$ and $\boldsymbol{P}$. In order to determine the full system of equations, additionally constitutive equations need to be developed. In fact, these equations establish a relation between the two functions $\boldsymbol{P}$ and $\boldsymbol{r}$ induced by material behavior, E.g. in context of elasticity the aim is to find constitutive equations of the form

$$\boldsymbol{P}(s, t) = \hat{\boldsymbol{P}}(\partial_s \boldsymbol{r}, s)\ . \tag{2.103}$$

Remark 2.15 (Geometrically exact beam formulation in the context of the general nonlinear continuum formulation). *The geometrically exact beam formulation analysed in Sec. 2.2 can be perceived as a one-dimensional continuum. In this remark we aim to work out the links to the general three-dimensional nonlinear continuum formulation (cf. [88, App. A]). For that purpose we restrict ourself to straight reference configurations defined by material coordinates* $\mathbf{s} = \begin{bmatrix} \xi_1 & \xi_2 & \xi_3 \end{bmatrix}^T$. *In Fig. 2.14 an illustration of the beam is depicted.*

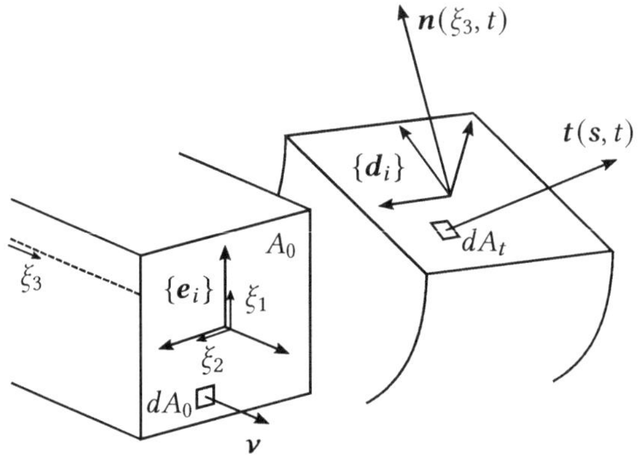

Figure 2.14.: Material (left) and spatial (right) configuration for the geometrically exact beam formulation.

The resultant contact force per unit reference length, acting on cross-section A_t *in spatial configuration can be established as*

$$\boldsymbol{n}(\xi_3, t) = \int_{A_0} \boldsymbol{t}(\mathbf{s}, t) \, \mathrm{d}A_0 \,. \tag{2.104}$$

Cauchy's theorem (2.95), *states that the stress vector* $\boldsymbol{t} : \Omega \mapsto \mathbb{R}^3$ *per unit reference area* A_0 *acting on the cross section* A_t *is determined by the second-order tensor field* $\boldsymbol{P} : \Omega \mapsto \mathbb{R}^{3,3}$ *and the normal vector* $\boldsymbol{\nu}$ *of the contact plane, i.e.* $\boldsymbol{t}(\mathbf{s}, t) = \boldsymbol{P}(\mathbf{s}, t) \cdot \boldsymbol{\nu}$. *Since, in context of the beam theory, the orientation of the contact surface is obviously always defined by the normal vector* $\boldsymbol{e}_3$, *the following applies*

$$\boldsymbol{P} \cdot \boldsymbol{e}_3 = (\boldsymbol{t}(\mathbf{s}, t, -\boldsymbol{e}_k) \otimes \boldsymbol{e}_k) \cdot \boldsymbol{e}_3 = (\boldsymbol{e}_k \cdot \boldsymbol{e}_3)\boldsymbol{t}(\mathbf{s}, t, -\boldsymbol{e}_k) = \boldsymbol{t}(\mathbf{s}, t, -\boldsymbol{e}_3) \,.$$

Therein, the definition of the first Piola-Kirchoff stress tensor field, i.e. $\boldsymbol{P} = \boldsymbol{t}(\mathbf{s}, t, -\boldsymbol{e}_k) \otimes \boldsymbol{e}_k$ *where* $k \in \{1, 2, 3\}$, *as well as the orthogonality condition* $\boldsymbol{e}_i \cdot \boldsymbol{e}_j = \delta_{ij}$ *have been exploited. Eq.* (2.104) *can then be reformulated as*

$$\boldsymbol{n}(\xi_3, t) = \int_{A_0} \boldsymbol{P} \cdot \boldsymbol{e}_3 \, \mathrm{d}A_0 = \int_{A_0} \boldsymbol{t}(\mathbf{s}, t, -\boldsymbol{e}_3) \, \mathrm{d}A_0 \,. \tag{2.105}$$

Taking into account that the divergence of the first Piola-Kirchoff stress tensor field is obtained as

$$\operatorname{div}_s \boldsymbol{P} = \partial_{\xi_k} P_{ik} \boldsymbol{e}_i \otimes \boldsymbol{e}_k = \partial_{\xi_k} \boldsymbol{t}(\mathbf{s}, t, -\boldsymbol{e}_k) \tag{2.106}$$

as well as making use of the local form of the balance of linear momentum (2.97), *the derivative of the resultant contact force* (2.104) *with respect to* ξ_3 *can be established as*

$$\begin{aligned}\partial_{\xi_3}\boldsymbol{n} &= \int_{A_0} \partial_{\xi_3}(\boldsymbol{t}(s,t,-\boldsymbol{e}_3))\ \mathrm{d}A_0 \\ &= \int_{A_0} \operatorname{div}\boldsymbol{P} - \left(\partial_{\xi_1}\boldsymbol{t}(s,t,-\boldsymbol{e}_1) + \partial_{\xi_2}\boldsymbol{t}(s,t,-\boldsymbol{e}_2)\right)\ \mathrm{d}A_0 \\ &= \int_{A_0} \rho\partial_t^2\boldsymbol{r} - \boldsymbol{b} - \left(\partial_{\xi_1}\boldsymbol{t}(s,t,-\boldsymbol{e}_1) + \partial_{\xi_2}\boldsymbol{t}(s,t,-\boldsymbol{e}_2)\right)\ \mathrm{d}A_0\ .\end{aligned}$$

Taking advantage of the divergence theorem

$$\int_A \partial_{\xi_i}\boldsymbol{t}(s,t,\boldsymbol{e}_i) = \int_{\partial A} \boldsymbol{t}(s,t,-\boldsymbol{e}_i)\cdot\boldsymbol{\nu}_i\ \mathrm{d}\Gamma \tag{2.107}$$

and introducing, in accordance to (2.94), *the resultant force*

$$\boldsymbol{f}(s,t) = \int_{\partial A} \boldsymbol{t}(s,t,-\boldsymbol{e}_i)\cdot\nu_i\ \mathrm{d}\Gamma + \int_A \boldsymbol{b}\ \mathrm{d}A\,, \tag{2.108}$$

the balance of linear momentum for the geometrically ecact beam is obtained as

$$\partial_{\xi_3}\boldsymbol{n} = (\rho A_0)(s)\partial_t^2\boldsymbol{r}(s,t) - \boldsymbol{f}(s,t)\ .$$

Compare with Eq. $(2.51)_1$. *Analogously, the balance of angular momentum of the geometrically exact beam formulation can be derived from the three-dimensional theory (cf. [88]).*

3. Sequential space-time integration

Initial boundary value problems that occur in non-linear structural dynamics are commonly solved by applying a sequential space-time discretization, cf. e.g. [34, 62]. The so-called method of lines is therefore typically based on a discretization in space by means of finite elements, followed by an appropriate discretization in time mostly based on finite differences. In Sec. 3.1, a brief survey of such sequential space-time integration methods for the initial boundary value problem at hand will be given. This will be done first in the context of the direct dynamics problem. For that purpose, the pure *Neumann* boundary problem is introduced before different possibilities of imposing *Dirichlet* boundary conditions in general, will be discussed afterwards. Based on this, in Sec. 3.2, the inverse dynamics problem will be introduced in the context of spatially discrete mechanical systems subjected to time-varying servo-constraints. A detailed analysis of such constrained problems aims to elaborate the fundamental distinctions between servo-constraints and classical contact-constraints. Consequences thereof, regarding the construction of numerically stable integration methods, will be addressed likewise before numerous selected examples will be given. This section partly reproduces [93].

3.1. Direct dynamics problem

Considering for the moment the formulation of the classical direct, pure *Neumann* problem[1] associated with the initial boundary value problem (1.1), then the aim is to find the overall motion of the system affected by external loads, neglecting any *Dirichlet* boundary conditions.

Multiplying Eq. (1.1) by sufficiently smooth test functions $\boldsymbol{w} : S \mapsto \mathbb{R}^d$ and subsequently integrating over the spatial domain S yields

$$\int_S \boldsymbol{w}(s) \cdot \boldsymbol{A}(s) \cdot \partial_t^2 \boldsymbol{x}(s,t) \, \mathrm{d}s - \int_S \boldsymbol{w}(s) \cdot \mathrm{div}_s \left(\boldsymbol{B}(s,t) \cdot \partial_s \boldsymbol{x}(s,t) \right) \mathrm{d}s = \int_S \boldsymbol{w}(s) \cdot \boldsymbol{C}(s,t) \, \mathrm{d}s \, . \tag{3.1}$$

[1] The pure *Neumann* problem is obtained by removing the displacement boundary conditions.

Applying integration by parts to the second integral on the left-hand side and taking into account the *Neumann* boundary conditions leads to an equivalent weak form of the boundary value problem at hand

$$\begin{aligned}\int_S \boldsymbol{w}(s) \cdot \boldsymbol{A}(s) \cdot \partial_t^2 \boldsymbol{x}(s,t) \, \mathrm{d}s + \int_S \partial_s \boldsymbol{w}(s) \cdot \boldsymbol{B}(s,t) \cdot \partial_s \boldsymbol{x}(s,t) \, \mathrm{d}s \\ = \int_S \boldsymbol{w}(s) \cdot \boldsymbol{C}(s,t) \, \mathrm{d}s - \boldsymbol{f}(t) \cdot \boldsymbol{w}(s,t)|_{(s,t) \in \partial S_f} \\ + \boldsymbol{\eta}(t) \cdot \boldsymbol{w}(s,t)|_{(s,t) \in \partial S_\eta} \ .\end{aligned} \tag{3.2}$$

This weak form can be discretized in space by applying suitable finite element approximations to the vector-valued test function $\boldsymbol{w}(s)$ and the trial function $\boldsymbol{x}(s,t)$. We confine our attention to piecewise linear approximations based on Lagrangian shape functions (cf. [51]). The corresponding interpolations read

$$\boldsymbol{w}^h(s) = \sum_{i=1}^{n_S} L_i(s) \boldsymbol{w}_i \quad \text{and} \quad \boldsymbol{x}^h(s,t) = \sum_{j=1}^{n_S} L_j(s) \boldsymbol{q}_j(t) \ , \tag{3.3}$$

where $L_i(s)$ are α-linear Lagrangian shape functions and $n_S \in \mathbb{N}$ denote the number of nodes. The associated nodal values $\boldsymbol{w}_i : \mathcal{T} \mapsto \mathbb{R}^d$ of the test functions are arbitrary, while $\boldsymbol{q}_j : \mathcal{T} \mapsto \mathbb{R}^d$ denotes the nodal position vectors at time $t \in \mathcal{T}$. That is, $\boldsymbol{q}_j(t) = \boldsymbol{x}^h(s_j, t)$.

Inserting the finite element approximations into weak form (3.2) yields the semi-discrete equations of motion

$$\boldsymbol{M} \, \mathrm{D}_t^2 \boldsymbol{q} = \boldsymbol{F}(\boldsymbol{q}, t) \ , \tag{3.4}$$

where

$$\boldsymbol{F}(\boldsymbol{q}, t) = \boldsymbol{f}^{\text{ext}}(t) - \boldsymbol{f}^{\text{int}}(\boldsymbol{q}) - \boldsymbol{B}_f^T \boldsymbol{f}(t) + \boldsymbol{B}_\eta^T \boldsymbol{\eta}(t) \ . \tag{3.5}$$

Here, $\boldsymbol{q} : \mathcal{T} \mapsto \mathbb{R}^{d \cdot n_S}$ is the nodal configuration vector that contains the nodal position vectors at time $t \in \mathcal{T}$ of the discrete problem at hand, i.e.

$$\boldsymbol{q}(t) = \begin{bmatrix} \boldsymbol{q}_1(t) \\ \vdots \\ \boldsymbol{q}_{p+1}(t) \end{bmatrix} . \tag{3.6}$$

Furthermore, the nodal contributions to the mass matrix $\boldsymbol{M}$, the internal force vector $\boldsymbol{f}^{\text{int}}(\boldsymbol{q})$ and the external force vector $\boldsymbol{f}^{\text{ext}}(t)$ are given by, respectively,

$$\begin{aligned} \boldsymbol{M}_{ij}(t) &= \boldsymbol{I}_d \int_S L_i(s) \boldsymbol{A}(s) L_j(s) \, \mathrm{d}s \ , \\ \boldsymbol{f}_i^{\text{int}}(\boldsymbol{q}, t) &= \int_S \mathrm{D}_s L_i(s) \boldsymbol{B}(s,t) \partial_s \boldsymbol{r}^h(s,t) \, \mathrm{d}s \ , \\ \boldsymbol{f}_i^{\text{ext}}(t) &= \int_S L_i(s) \boldsymbol{C}(s,t) \, \mathrm{d}s \ , \end{aligned} \tag{3.7}$$

where $\boldsymbol{I}_d$ is the $d \times d$ identity matrix. The matrix $\boldsymbol{B}_f$ in (3.4) is of Boolean type and essentially links the force $\boldsymbol{f}(t)$ to the nodes lying on $\partial\Omega_f$, accounting for the virtual work contribution $\delta\boldsymbol{q} \cdot \boldsymbol{B}_f^T \boldsymbol{f}(t) = \boldsymbol{w}_1 \cdot \boldsymbol{f}(t)$ emerging from the right-hand side of weak form (3.2). Here, $\delta\boldsymbol{q}$ contains the nodal values $\boldsymbol{w}_i$ in analogy to (3.6).

The semi-discrete equations of motion (3.4) constitute a system of non-linear ordinary differential equations of second-order. In the standard forward dynamics problem, the forcing terms $\boldsymbol{f}_{\text{ext}}(t)$, $\boldsymbol{f}(t)$ and $\boldsymbol{\eta}(t)$ are prescribed functions of time and (3.4) can be solved by applying common time-stepping schemes.

Example 3.1 (Linear elastic bar). *In the case of the linear-elastic bar (Ex. 2.5), the semi-discrete equations of motion* (3.4) *boil down to a system of ordinary differential equations with constant coefficients. Confining to piecewise linear approximations, the nodal contributions to Eq.* (3.5) *can be evaluated to be*

$$F_i(t) = K_{ij} u_j(t) - 0 - \delta_{i1} f(t) + \delta_{i,p+1} \eta(t) \tag{3.8}$$

where for each element

$$K_{ij}^{(e)} = \frac{EA}{h_s^{(e)}} \left(\delta_{ij} - \delta_{i+1,j} - \delta_{i-1,j}\right)$$

holds. By assuming lumped masses, the nodal contributions to the mass matrix read

$$M_{ij}^{(e)} = \rho A \frac{h_s^{(e)}}{2} = \frac{m^{(e)}}{2} \delta_{ij} \ .$$

For the more general case of as deriving a consistent mass matrix, the contributions to the mass matrix are obtained as

$$M_{ij}^{(e)} = \rho A \frac{h_s^{(e)}}{6} \left(2\delta_{ij} + \delta_{i+1,j} + \delta_{i-1,j}\right)$$

again by confining to piecewise linear Lagrangian shape functions.

Remark 3.2 (Additional point mass). *If an additional point mass M is attached to the right end of the string (Rem. 2.2), $\boldsymbol{\eta}(t) = M\left(\boldsymbol{g} - \partial_t^2 \boldsymbol{r}(L,t)\right)$ has to be taken into account in weak form* (3.2)*. Correspondingly, in the semi-discrete formulation, the following entries of mass matrix* $(3.7)_1$ *and external load vector* $(3.7)_3$ *need to be modified according to*

$$\begin{aligned} \boldsymbol{M}_{p+1,p+1} &\leftarrow \boldsymbol{M}_{p+1,p+1} + M\boldsymbol{I}_d \ , \\ \boldsymbol{f}_{p+1}^{\text{ext}} &\leftarrow \boldsymbol{f}_{p+1}^{\text{ext}} + M\boldsymbol{g} \ . \end{aligned}$$

Dirichlet Boundary conditions. Boundary conditions pertaining to the configuration space $Q = \mathbb{R}^{d \cdot n_S}$ of the underlying mechanical system, i.e.

$$\boldsymbol{x}|_{(s,t) \in \partial\Omega_D} = \bar{\boldsymbol{x}}(s,t) \ , \tag{3.9}$$

are called *Dirichlet* boundary conditions. Classically, such conditions are taken into account by choosing suitable spaces for the test functions, i.e.

$$W = \{\boldsymbol{w} : \Omega \mapsto \mathbb{R}^d \mid \boldsymbol{w} = \boldsymbol{0} \text{ on } \partial S_D \times \mathcal{T}\} \ . \tag{3.10}$$

The system at hand can then be formulated in a generalized, minimal configuration space, i.e. $\boldsymbol{q} : \mathcal{T} \mapsto \mathbb{R}^{d\cdot(n_S - n_D)}$, representing the nodal configuration vector that contains the nodal position vectors, except for the nodes lying on the Dirichlet boundary ∂S_D, at time $t \in \mathcal{T}$ of the discrete problem at hand. The solution is then governed by ordinary differential equations of the form

$$\boldsymbol{M} \cdot \mathrm{D}_t^2 \boldsymbol{q} + \boldsymbol{F} = 0 \ . \tag{3.11}$$

Alternatively, boundary conditions of the form (3.9) might be ensured by imposing feasible geometric constraints to the configuration space Q. The underlying system can then be formulated in terms of redundant coordinates $\boldsymbol{q}$ on a specified constraint manifold $C = \{\boldsymbol{q} : \boldsymbol{g}(\boldsymbol{q}) = 0\}$, such that the motion is governed by a semi-explicit system of differential-algebraic equations[2] given by

$$\begin{aligned} \boldsymbol{M}\,\mathrm{D}_t^2 \boldsymbol{q} + \boldsymbol{F}(\mathrm{D}_t\boldsymbol{q}, \boldsymbol{q}, t) + \boldsymbol{G}^T \boldsymbol{\lambda}(t) &= \boldsymbol{0} \ , \\ \boldsymbol{g}(t, \boldsymbol{q}) = \boldsymbol{G}\boldsymbol{q} - \boldsymbol{\gamma}(t) &= \boldsymbol{0} \ , \end{aligned} \tag{3.12}$$

where $\boldsymbol{\lambda} : \mathcal{T} \mapsto \mathbb{R}^d$ are the Lagrange multipliers enforcing the given constraints $\boldsymbol{g} : \mathcal{T} \times Q \mapsto \mathbb{R}^{d\cdot n_D}$. Furthermore, the constraint jacobian $\boldsymbol{G} = \partial_{\boldsymbol{q}} \boldsymbol{g}(t, \boldsymbol{q})$ has been introduced in Eq. (3.12). For more details on constrained mechanical systems, see Rem. 3.12, Ex. 3.13 and references therein. By introducing the velocity $\boldsymbol{v} \equiv \mathrm{D}_t \boldsymbol{q}$, the system (3.12) might be reformulated as a first-order system

$$\begin{aligned} \mathrm{D}_t \boldsymbol{v} &= \mathcal{F}_1(\boldsymbol{v}, \boldsymbol{q}, \boldsymbol{\lambda}) = \boldsymbol{M}^{-1}(\boldsymbol{f}(\boldsymbol{v}, \boldsymbol{q}) - \boldsymbol{G}^T \lambda) \ , \\ \mathrm{D}_t \boldsymbol{q} &= \mathcal{F}_2(\boldsymbol{v}, \boldsymbol{q}) = \boldsymbol{v} \ , \\ \boldsymbol{0} &= \mathcal{F}_3(\boldsymbol{q}) = \boldsymbol{g}(\boldsymbol{q}) \ , \end{aligned} \tag{3.13}$$

with non-singular

$$\partial_{\boldsymbol{q}} \mathcal{F}_3 \cdot \partial_{\boldsymbol{v}} \mathcal{F}_2 \cdot \partial_\lambda \mathcal{F}_1 = -\boldsymbol{G}\boldsymbol{M}^{-1}\boldsymbol{G}^T \ . \tag{3.14}$$

Then Eq. (3.13) can be identified as a system of differential-algebraic equations in Hessenberg form of index $\nu_d = 3$ (cf. Def. 3.7 and Def. 3.6 for $\boldsymbol{z} = \begin{bmatrix} \boldsymbol{v} & \boldsymbol{q} & \boldsymbol{\lambda} \end{bmatrix}^T$ and references therein).

In the case of ordinary contact constraints, the constraint forces $\boldsymbol{G}^T \lambda$ are ideal-orthogonal to the contact constraint manifold $C = \{\boldsymbol{q} : \boldsymbol{g}(\boldsymbol{q}) = 0\}$ and $rank(\boldsymbol{G}\boldsymbol{M}^{-1}\boldsymbol{G}^T) = d \cdot n_S$ applies.

[2] Also frequently referred to as *algebro-differential equations*, *implicit differential equations* or *singular systems* ([63]).

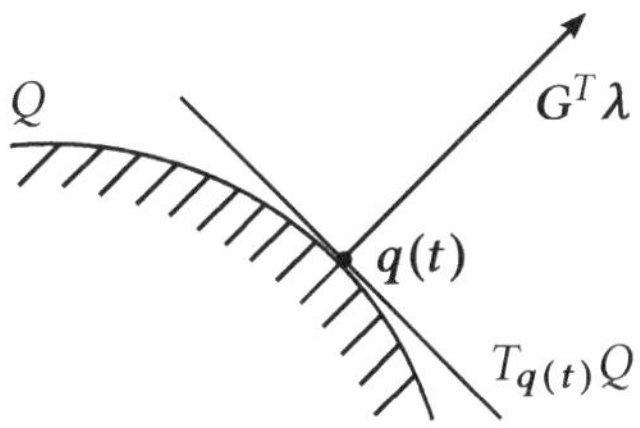

Figure 3.1.: Ideal-orthogonal contact constraint realization.

The semi discrete equations of motion can then be solved by subsequently integrating (3.12) in time by applying suitable finite difference schemes (cf. e.g, [66]). It should be kept in mind, that suitable stabilization techniques such as the *Gear-Gupta-Leimkuhler* stabilization (cf. Rem. 3.8) or the *Baumgarte* stabilization (cf. Rem. 3.9) are mandatory for temporal integration with error controlled variable time step size ([10]).

In Ex. 3.10 and Ex. 3.11, two classical examples of inherent spatially discrete systems that align with system (3.12) are presented.

Definition 3.3 (Constraints). *Geometric constraints pertaining the configuration space as given by*

$$\boldsymbol{g}(t,\boldsymbol{q}) = \boldsymbol{0} \tag{3.15}$$

are termed holonomic. Otherwise, the constraints are termed non-holonomic. Examples of non-holonomic constraints are, e.g.

(i) Inequality constraints of the form

$$\boldsymbol{g}(t,\boldsymbol{q}) \leq \boldsymbol{0}\ .$$

(ii) Constraints in differential but non-integrable form

$$\boldsymbol{g}(t,\boldsymbol{q},\mathrm{D}_t\boldsymbol{q}) = \boldsymbol{0}\ . \tag{3.16}$$

I.e. there exists no function $\boldsymbol{\mathcal{G}}(t,\boldsymbol{q})$ such that $\partial_t\boldsymbol{\mathcal{G}}(t,\boldsymbol{q}) = \boldsymbol{g}(t,\boldsymbol{q},\mathrm{D}_t\boldsymbol{q})$. Any integration of constraint (3.16) *will not lead to a constraint in form of* (3.15)*. An example of a constraint in differential representation that is integrable is a non-sliding coin on a line (cf. Ex. 3.4). In contrast to that, the same constraint forcing the coin to roll without sliding on a plane is non-holonomic (cf. Ex. 3.5).*

Furthermore, a constraint is called rheonomic if the constraint depends explicitly on the time variable, i.e. $\partial_t\boldsymbol{g} \neq \boldsymbol{0}$. Otherwise, i.e. $\partial_t\boldsymbol{g} = \boldsymbol{0}$, the constraint is called scleronomic. For more details on the classification of constraints see [44, pp. 12-16] and [70, pp. 236-237].

Example 3.4 (Coin on a line: Holonomic constraint). *The kinematics of a circular coin with radius r rolling on a line without sliding is uniquely defined by the coordinates* $\boldsymbol{q} = \begin{bmatrix} x & \varphi \end{bmatrix}^T$ *in addition to the scleronomic constraint in differential form as given by*

$$g(\boldsymbol{q}, \mathrm{D}_t\boldsymbol{q}) = \mathrm{D}_t x - r\,\mathrm{D}_t\varphi = 0 \tag{3.17}$$

Since a function $\mathcal{G}(\boldsymbol{q}) = x - r\varphi = const.$ *exists, the constraint* (3.17) *is holonomic.*

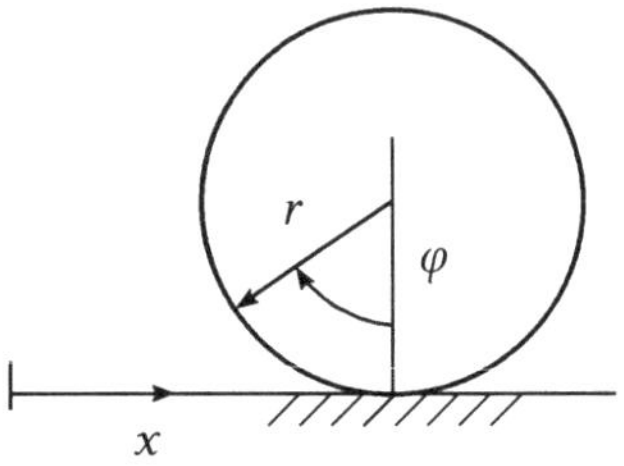

Figure 3.2.: Holonomic constraint: Non-sliding coin on a line.

Example 3.5 (Coin on a plane: Non-holonomic constraint). *Reconsidering Ex. 3.4 for the case of a rolling coin on a* plane, *the kinematics of the coin can be described in terms of its location* $\boldsymbol{q}_1 = \begin{bmatrix} x & y \end{bmatrix}$ *and its orientation* $\boldsymbol{q}_2 = \begin{bmatrix} \varphi & \Theta \end{bmatrix}$. *The condition of pure rolling, e.g. setting the orientation and the location of the coin into relation, can be ensured by imposing the following scleronomic constraints*

$$g(\boldsymbol{q}, \mathrm{D}_t\boldsymbol{q}) = \boldsymbol{v} - r(\mathrm{D}_t\varphi)\boldsymbol{n}(\Theta) = \begin{bmatrix} \mathrm{D}_t x \\ \mathrm{D}_t y \end{bmatrix} - r\,\mathrm{D}_t\varphi \begin{bmatrix} \sin\Theta \\ \cos\Theta \end{bmatrix} = \boldsymbol{0} \tag{3.18}$$

depending on the coordinates $\boldsymbol{q} = \begin{bmatrix} \boldsymbol{q}_1 & \boldsymbol{q}_2 \end{bmatrix}^T$. *If* (3.18) *would be holonomic, a functional relation between* $\boldsymbol{q}_1$ *and* $\boldsymbol{q}_2$ *could be established, hence* (3.18) *could be uniquely solved by integration. This is obiously not the case. Various maneuvers of the coin in a time interval* $\mathcal{T} = [0, T]$ *are conceivable such that one coordinate can be chosen independently at time* T. *Consider exemplarily the maneuver of tracing a circle, such that* $x(0) = x(T)$, $y(0) = y(T)$ *and* $\Theta(0) = \Theta(T)$. *Obviously* $\varphi(0) \neq \varphi(T)$ *depending on the chosen radius of the circle.*

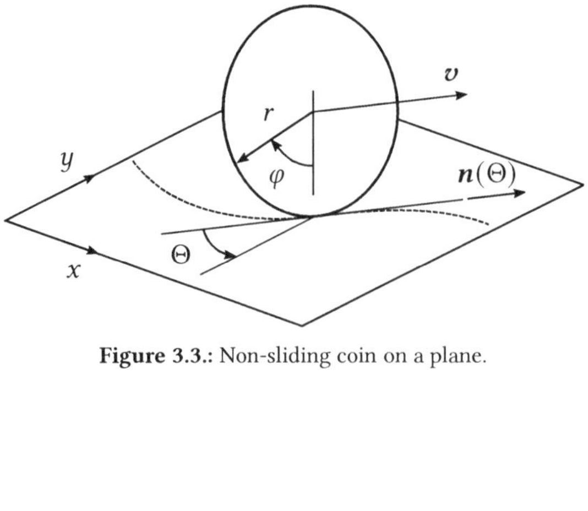

Figure 3.3.: Non-sliding coin on a plane.

Definition 3.6 (Differential-algebraic equations in Hessenberg form). *Differential-algebraic equations in the form of*

$$\begin{aligned} \mathrm{D}_t z_1 &= \mathcal{F}_1(z_1, ..., z_\nu) \\ \mathrm{D}_t z_2 &= \mathcal{F}_2(z_1, ..., z_{\nu-1}) \\ \mathrm{D}_t z_3 &= \mathcal{F}_3(z_2, ..., z_{\nu-1}) \\ &\vdots \\ \mathrm{D}_t z_{\nu-1} &= \mathcal{F}_{\nu-1}(z_{\nu-2}, ..., z_{\nu-1}) \\ 0 &= \mathcal{F}_\nu(z_{\nu-1}) \end{aligned}$$

with a nonsingular

$$\partial_{z_{\nu-1}}\mathcal{F}_\nu \cdot \partial_{z_{\nu-2}}\mathcal{F}_{\nu-1} \cdots \partial_{z_1}\mathcal{F}_2 \cdots \partial_{z_\nu}\mathcal{F}_2$$

are called DAEs in Hessenberg form of differentiation index ν_d (cf. [63, p.172]).

Definition 3.7 (Differentiation index). *The differentiation index ν_d of a general differential-algebraic equation*

$$\mathcal{F}(z', z, t) = 0 \tag{3.19}$$

along with a singular $\partial_{z'}\mathcal{F}$, is defined by the minimum number $m \in \mathbb{N}$ of derivatives

$$\begin{aligned} \mathrm{D}_t\mathcal{F}(z', z, t) &= (\partial_{z'}\mathcal{F})\,\mathrm{D}_t^2 z + (\partial_z)(\mathrm{D}_t z) + \partial_t\mathcal{F} = 0 \\ \mathrm{D}_t^2\mathcal{F}(z', z, t) &= (\partial_{z'}\mathcal{F})\,\mathrm{D}_t^3 z + \cdots = 0 \\ &\vdots \\ \mathrm{D}_t^m\mathcal{F}(z', z, t) &= (\partial_{z'}\mathcal{F})\,\mathrm{D}_t^{s+1} z + \cdots = 0 \end{aligned} \tag{3.20}$$

that need to be considered, such that Eq. (3.19) can be solved for $z'(t) = \mathrm{D}_t z(t)$. This definition of the differentiation index of general differential-algebraic equations can be traced back to C.W. Gear (cf. [40] and [41]). For a comprehensive study of the differentiation index see [63, pp.96-113].

Remark 3.8 (Gear-Gupta-Leimkuhler stabilization). *The original idea of the Gear-Gupta-Leimkuhler stabilization is to minimally extend the differential-algebraic system (3.13) by temporal derivatives of the constraint equation. These additional constraints are enforced by additional Lagrange multipliers. For Hessenberg systems with differentiation index $\nu_d = 3$, i.e. mechanical systems subjected to holonomic ideal-orthogonal contact constraints, the stabilized system in terms of Gear-Gupta-Leimkuhler has differentiation index $\nu_d = 2$ and reads*

$$\begin{aligned} \mathrm{D}_t \boldsymbol{q} &= \boldsymbol{v} - \boldsymbol{G}^T \boldsymbol{\mu}\,, \\ \boldsymbol{M}\,\mathrm{D}_t \boldsymbol{v} &= \boldsymbol{F} - \boldsymbol{G}^T \boldsymbol{\lambda}\,, \\ \boldsymbol{0} &= \boldsymbol{g}\,, \\ \boldsymbol{0} &= \partial_{\boldsymbol{q}}\boldsymbol{g}\,\mathrm{D}_t\boldsymbol{q} = \boldsymbol{G}\boldsymbol{v}\,. \end{aligned}$$

This stabilization technique can be traced back to [42].

Remark 3.9 (Baumgarte stabilization). *In principle, the second time derivative of the constraint equation*

$$\mathrm{D}_t^2 \boldsymbol{g}(\boldsymbol{q}) = \boldsymbol{0} \tag{3.21}$$

could be used to eliminate the Lagrange multipliers to integrate the differential-algebraic system of equation (3.19) *with differentiation index* $\nu_d = 3$ *numerically. Unfortunately the resulting system of ordinary differential equations turns out to be be unstable in the sense of Ljapunov and drift-off phenomena may occur. To avoid this, the following stabilization method is suggested by* J. Baumgarte *in [15]. Instead of using only the differentiated constraint equation* (3.21), *a combination of the original and differentiated constraint equations may be taken into account, i.e.*

$$\mathrm{D}_t^2 \boldsymbol{g} + 2\alpha\, \mathrm{D}_t \boldsymbol{g} + \beta^2 \boldsymbol{g} = \boldsymbol{0}, \quad \alpha > 0 \tag{3.22}$$

in the case of holonomic and

$$\mathrm{D}_t \boldsymbol{g} + \gamma \boldsymbol{g} = \boldsymbol{0}, \quad \gamma > 0 \tag{3.23}$$

in the case of non-holonomic constraints. In Eq. (3.22) *and Eq.* (3.23), *the non-negative and constant parameters* α, β *and* γ *have been introduced.*

Example 3.10 (Planar mathematical pendulum). *The planar mathematical pendulum is an important prototype of a constrained mechanical system. Its virtue lies in encompassing the mathematical structure without being distracted by superfluous complexity. A particle in a gravitational field* $g = 1$ *with mass* $m = 1$ *is forced to stay on a circle by a rigid and weightless rod with length* $l = 1$ *attached to an inertial frame. Its motion is governed either by ordinary differential or differential-algebraic equations, depending on whether redundant or generalized coordinates are chosen. Both formulations will be addressed subsequently.*

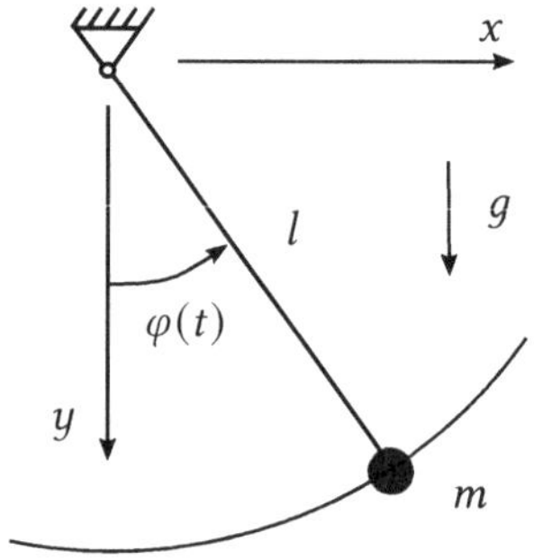

Figure 3.4.: Illustration of a planar mathematical pendulum.

(i) *Redundant coordinates. By choosing cartesian coordinates* $\boldsymbol{q} = \begin{bmatrix} x & y \end{bmatrix}^T$, *the Lagrangian of the planar mathematical pendulum can be established as*

$$L(t, \boldsymbol{q}, \mathrm{D}_t \boldsymbol{q}) = \frac{1}{2} \mathrm{D}_t \boldsymbol{q}^T \mathrm{D}_t \boldsymbol{q} + y = \frac{1}{2}((\mathrm{D}_t x)^2 + (\mathrm{D}_t y)^2) + y\,.$$

Together with the scleronomic holonomic constraint

$$g(\boldsymbol{q}) = \frac{1}{2}\left(\boldsymbol{q}^T\boldsymbol{q} - 1\right) = \frac{1}{2}\left(x^2 + y^2 - 1\right) = 0 \tag{3.24}$$

and its corresponding Jacobian

$$\boldsymbol{G} = \begin{bmatrix} x & y \end{bmatrix} = \boldsymbol{q}\,, \tag{3.25}$$

the motion of the mathematical pendulum is governed by the Euler-Lagrange equations for constrained mechanical systems (cf. Rem. 3.12, Eq. (3.38)*) taking the form of a system of differential-algebraic equations as given by*

$$\begin{aligned} \mathrm{D}_t^2 x &= -\lambda x\,, \\ \mathrm{D}_t^2 y &= -\lambda y + 1\,, \\ g(x,y) &= 0\,. \end{aligned} \tag{3.26}$$

By introducing

$$\boldsymbol{M} = \boldsymbol{I}, \qquad \boldsymbol{F} = \begin{bmatrix} 0 & 1 \end{bmatrix}^T \quad \textit{and} \quad \boldsymbol{G} = \boldsymbol{q}$$

the problem at hand aligns with system (3.12)*. The system of equations* (3.26) *is sometimes referred to as Lagrange's equations of the first kind (cf. Rem. 3.12, Ex. 3.13 and references therein). By introducing the velocity* $\boldsymbol{v} \equiv \mathrm{D}_t\boldsymbol{q}$ *a Hessenberg form with differentiation index* $\nu_d = 3$ *can be identified (cf. Def. 3.6).*

This might be proven by transforming the differential-algebraic system of equations into an equivalent system of ordinary differential equations by using derivatives of the system and parts of it. In the context of the mathematical pendulum, the differential part of system (3.26) *might be inserted into the second time derivative of the scleronomic holonomic constraint equation* (3.24) *as given by*

$$\mathrm{D}_t^2 g(r(t)) = x\,\mathrm{D}_t^2 x + (\mathrm{D}_t x)^2 + y\,\mathrm{D}_t^2 y + (\mathrm{D}_t y)^2 = 0\,.$$

This leads to a relation for the Lagrange multiplier given by

$$\lambda = (\mathrm{D}_t x)^2 + (\mathrm{D}_t y)^2 + y\,. \tag{3.27}$$

A further differentiation of Eq. (3.27) *with respect to time*

$$\begin{aligned} \mathrm{D}_t\lambda &= 2(\mathrm{D}_t x)(\mathrm{D}_t^2 x) + 2(\mathrm{D}_t y)(\mathrm{D}_t^2 y) + \mathrm{D}_t y \\ &= 2\lambda\,(x\,\mathrm{D}_t x + y\,\mathrm{D}_t y) + 3\,\mathrm{D}_t y \\ &= 3\,\mathrm{D}_t y \end{aligned}$$

gives rise to the following system of ordinary differential equations (cf. [42] and example therein)

$$\begin{aligned} \mathrm{D}_t\boldsymbol{q} - \boldsymbol{v} &= \boldsymbol{0}\,, \\ \mathrm{D}_t\boldsymbol{v} + \lambda\boldsymbol{q} - \boldsymbol{F} &= 0\,, \\ \mathrm{D}_t\lambda - 3\,\mathrm{D}_t y &= 0\,. \end{aligned}$$

Thus, the differential-algebraic system of equations (3.26) *could be transformed into a system of ordinary differential equations by differentiating three times in total. Hence, the differential-algebraic system of equations indeed has a differentiation index* $\nu_d = 3$.

In Fig. 3.5, an illustration of an ideal-orthogonal constraint representation is depicted. Therein $\boldsymbol{G}^T\lambda$*, that enforces the contact constraint, is ideal-orthogonal to the constraint manifold* $C = \{\boldsymbol{r} : g(\boldsymbol{q}) = \frac{1}{2}\left(\boldsymbol{q}^t\boldsymbol{q} - \gamma\right) = 0\}$.

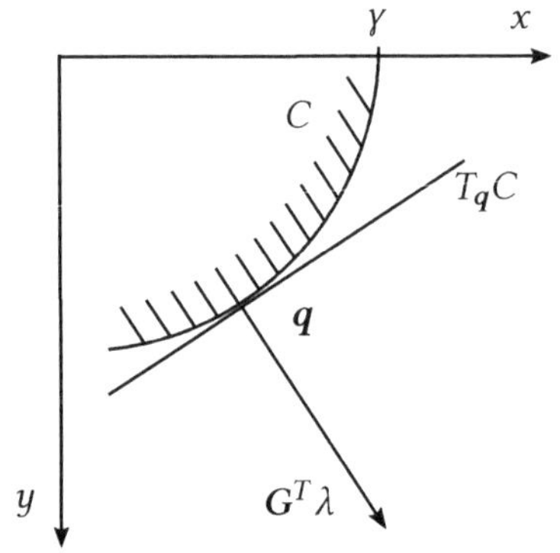

Figure 3.5.: Ideal-orthogonal constraint representation in context of the mathematical pendulum.

(ii) *Generalized coordinates. Alternatively, the pendulum can be formulated in terms of a generalized coordinate* φ. *Therefore, the differential-algebraic system of equations* (3.26) *along with* (3.24) *can be transformed into one single ordinary differential equation. Therefore a suitable coordinate transformation* $\mathcal{F}(\varphi) : \mathbb{R} \mapsto \mathbb{R}^2$ *such that* $\boldsymbol{q} \mapsto \mathcal{F}^{-1}(\varphi)$ *need to be identified merely. For instance, establishing the transformation*

$$\mathcal{F}(\varphi) = \begin{bmatrix} \sin(\varphi) \\ \cos(\varphi) \end{bmatrix} \tag{3.28}$$

a relation for the Lagrange multiplier, as given by

$$\lambda = \cos^2(\varphi)(\mathrm{D}_t\varphi)^2 + \sin^2(\varphi)(\mathrm{D}_t\varphi)^2 + \cos(\varphi) = (\mathrm{D}_t\varphi)^2 + \cos(\varphi)$$

is obtained. Inserting the transformation (3.28), *along with its first and second time derivative given by*

$$\mathrm{D}_t\mathcal{F}(\varphi) = \begin{bmatrix} \cos(\varphi) \\ -\sin(\varphi) \end{bmatrix} \mathrm{D}_t\varphi \tag{3.29}$$

and

$$\mathrm{D}_t^2\mathcal{F}(\varphi) = \begin{bmatrix} -\sin(\varphi) \\ -\cos(\varphi) \end{bmatrix} \mathrm{D}_t\varphi + \begin{bmatrix} \cos(\varphi) \\ -\sin(\varphi) \end{bmatrix} \mathrm{D}_t^2\varphi = \left(\mathrm{D}_t\mathcal{F}(\varphi) - \mathcal{F}(\varphi)\right)\mathrm{D}_t\varphi\ ,$$

respectively, into equation (3.26) *gives rise to*

$$-\sin(\varphi)(\mathrm{D}_t\varphi)^2 + \cos(\varphi)(\mathrm{D}_t^2\varphi)^2 = -\left((\mathrm{D}_t\varphi)^2 + \cos(\varphi)\right)\sin(\varphi) \ .$$

Eventually, the system of differential-algebraic equations (3.26) *can finally be transformed into one single ordinary differential equation*

$$\mathrm{D}_t^2\varphi(t) + \sin(\varphi) = 0$$

governing the motion of the planar pendulum in terms of a single generalized coordinate.

Example 3.11 (Overhead crane - direct dynamics problem). *Besides the mathematical pendulum, analyzed in Ex. 3.10 as a prototype of constrained mechanical systems, the planar model of an overhead crane will be considered below (cf. Fig. 3.6 for an illustration thereof). This model serves as an important archetype for the inverse dynamics of underactuated mechanical systems. Therefore, it is first introduced generally in context of the direct dynamics problem.*

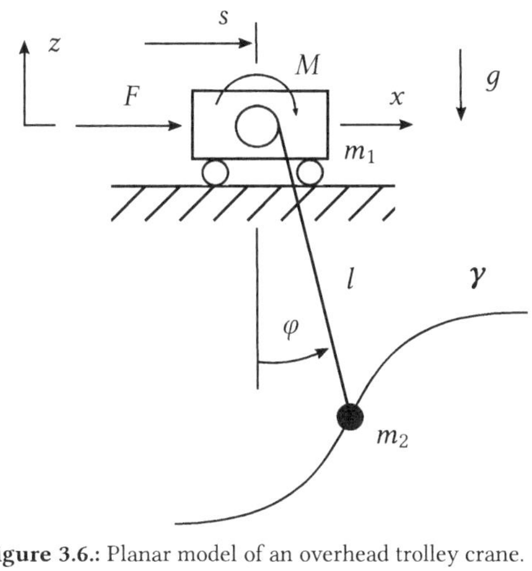

Figure 3.6.: Planar model of an overhead trolley crane.

A load with mass $m_2 = 1$ is attached to a trolley with mass $m_1 = 1$ via an inextensible and massless rope. The rope itself is connected to the trolley using a pulley with radius $r = 1$ and moment of inertia $J = 1$, such that the length $l : \mathcal{T} \mapsto \mathbb{R}$ of the rope is adjustable by exerting a torque $M : \mathcal{T} \mapsto \mathbb{R}$ on the pulley. Furthermore, the trolley is capable of being maneuvered along a path $s \in \mathbb{R}$ caused by a force $F : \mathcal{T} \mapsto \mathbb{R}$ acting horizontally on the trolley. The actuation $\boldsymbol{F} = \begin{bmatrix} F & M \end{bmatrix}^T$ causes a motion of the load in the x-z-plane. Without losing generality, the gravitational constant is assumed to be $g = 1$.

Similarly to the mathematical pendulum discussed in Ex. 3.10, the motion of the crane can be described either in redundant or generalized coordinates. Both formulations will be addressed below.

(i) *Redundant coordinates. In case of choosing redundant coordinates* $\boldsymbol{q} = \begin{bmatrix} s & l & x & z \end{bmatrix}^T$, *the Lagrangian of the system at hand is obtained as*

$$L = \frac{1}{2}\left[(\mathrm{D}_t s)^2 + (\mathrm{D}_t \varphi)^2 + (\mathrm{D}_t x)^2 + (\mathrm{D}_t z)^2\right] + z\,.$$

Additionally a scleronomic holonomic constraint given by

$$g(\boldsymbol{q}) = \frac{1}{2}((x-s)^2 + z^2 - l^2) = 0\,, \tag{3.30}$$

need to be taken into account. Then the Euler-Lagrange equations governing the motion of the crane are obtained as

$$\begin{aligned} &\mathrm{D}_t^2 s + \lambda(s-x) - F = 0\,,\\ &\mathrm{D}_t^2 l - \lambda l - M = 0\,,\\ &\mathrm{D}_t^2 x + \lambda(x-s) = 0\,,\\ &\mathrm{D}_t^2 z + \lambda z - 1 = 0\,,\\ &g(s,l,x,z) = 0 \end{aligned} \tag{3.31}$$

See Rem. 3.12 and references therein for more details on the Euler-Lagrange equations of constrained mechanical systems. The system of differential-algebraic equations (3.31) *aligns with system* (3.12) *considering the constraint Jacobian*

$$\boldsymbol{G} = \partial_{\boldsymbol{q}} g(\boldsymbol{q}) = \begin{bmatrix} -(x-s) & -l & (x-s) & z \end{bmatrix}^T$$

as well as $\boldsymbol{F} = \begin{bmatrix} F & M & 0 & -1 \end{bmatrix}^T$ *and* $\boldsymbol{M} = \boldsymbol{I}$. *Consequently, system* (3.31) *can be identified as a differential-algebraic system of equations in Hessenberg form of index* $\nu_d = 3$. *The constraint realization is ideal-orthogonal.*

(ii) *Generalized coordinates. Instead of describing the motion of the overhead crane in redundant coordinates, generalized coordinates might be chosen alternatively. For that purpose, a function* $\varphi(\boldsymbol{q})$ *can obviously be found, such that* $x = s + l\sin\varphi$ *and* $z = l\cos\varphi$ *holds. The motion of the overhead crane sketched in Fig. 3.6 can then be reformulated in terms of the generalized coordinates* $\boldsymbol{q} = \begin{bmatrix} s & l & \varphi \end{bmatrix}^T$. *The corresponding Lagrangian then reads*

$$L(\mathrm{D}_t\boldsymbol{q}, \boldsymbol{q}) = \frac{1}{2}\left[(\mathrm{D}_t s)^2 + (\mathrm{D}_t l)^2 + (\mathrm{D}_t(s + l\sin\varphi))^2 + (\mathrm{D}_t(l\cos\varphi))^2\right]. \tag{3.32}$$

In Eq. (3.32) *a non-sliding contact of the rope and the pulley, i.e.* $\mathrm{D}_t l = r\,\mathrm{D}_t\varphi$, *has been assumed. The motion of the crane is then governed by a system of ordinary differential equations in the form of* (3.11) *with*

$$\boldsymbol{M} = \begin{bmatrix} 2 & \sin\varphi & l\cos\varphi \\ \sin\varphi & 2 & 0 \\ l\cos\varphi & 0 & l^2 \end{bmatrix}$$

and

$$F = \begin{bmatrix} 2(\mathrm{D}_t l)(\mathrm{D}_t \varphi)\cos\varphi - l^2(\mathrm{D}_t\varphi)^2 \sin\varphi \\ -l(\mathrm{D}_t\varphi)^2 - \cos\varphi \\ 2l(\mathrm{D}_t l)(\mathrm{D}_t\varphi) - l\sin\varphi \end{bmatrix} - \begin{bmatrix} F \\ M \\ 0 \end{bmatrix}$$

Remark 3.12 (Constrained mechanical systems). *The equations of motion, postulated for several (constrained) mechanical systems will be derived in more detail subsequently. The variational problem with (finite) constraints based on Hamiltons principle can be stated as follows: Find a function* $\boldsymbol{q} : \mathcal{T} \mapsto \mathbb{R}^n$, *that extremizes the action*

$$S = \int_{t_0}^{t_1} L(t, \boldsymbol{q}, \mathrm{D}_t \boldsymbol{q})\ \mathrm{D}t\ , \tag{3.33}$$

compared to all varied functions

$$\bar{\boldsymbol{q}} = \boldsymbol{q} + \delta\boldsymbol{q} = \boldsymbol{q} + \alpha\boldsymbol{\eta}(t)\ , \tag{3.34}$$

by additionally complying to holonomic[3] *constraints of the form*

$$\boldsymbol{g}(t, \boldsymbol{q}) = \boldsymbol{0}\ . \tag{3.35}$$

In (3.34), *the function* $\delta\boldsymbol{q}(t) = \alpha\boldsymbol{\eta}(t)$ *is called the variation of the desired function* $\boldsymbol{q}(t)$ *that extremizes the action* (3.33). *Thereby, the varied function* $\bar{\boldsymbol{q}}(t)$ *is supposed to lie in a sufficiently close neighbourhood*[4] *of the solution* $\boldsymbol{q}(t)$. *Thus, the action* (3.33) *might be perceived as a function depending on the coefficient* α

$$\Phi(\alpha) = \int_{t_0}^{t_1} L(t, \bar{\boldsymbol{q}}(\alpha), \mathrm{D}_t\bar{\boldsymbol{q}}(\alpha))\ \mathrm{d}t \tag{3.36}$$

such that (3.36) *is extremized for* $\alpha = 0$ *compared to all sufficiently small* α. *Hence,*

$$\mathrm{D}_\alpha \Phi(\alpha)|_{\alpha=0} = 0\ .$$

Concurrently, the constraints (3.35) *must not be violated. According to the Lagrange multiplier theorem (cf. e.g. [70, pp. 234-235]) there exist a function* $\boldsymbol{\lambda} : \mathcal{T} \mapsto \mathbb{R}^m$ *such that the Euler-Lagrange equations*

$$-[L]_{\boldsymbol{q}} = \mathrm{D}_t(\partial_{\boldsymbol{q}} L) - \partial_{\boldsymbol{q}} L + \boldsymbol{\lambda}^T \partial_{\boldsymbol{q}} \boldsymbol{g} = \boldsymbol{0} \tag{3.37}$$

constitute necessary conditions to the function $\boldsymbol{q}(t)$ *extremizing the action* (3.33). *See Ex. 3.13 for the case of* $n = 2$ *independent functions subjected to holonomic constraints as well as references therein, covering the more general case of* $n > 2$ *functions subjected to non-holonomic constraints. Note that the Jacobian* $\mathbf{G} = \partial_{\boldsymbol{q}}\boldsymbol{g}(t, \boldsymbol{q})$ *is supposed to be non-singular. Eq.* (3.37)

[3] The correlations, that are made subsequently are extendable to the more general case of non-holonomic constraints (cf. [48],[27],[46]).

[4] For that purpose the scalar coefficient α needs to be chosen small enough

along with the constraints (3.35) *are frequently referred to as 'Lagrange's equations of motion of first kind'.*

By introducing an augmented Lagrangian in form of $L^* = L - \boldsymbol{\lambda}^T \boldsymbol{g}$*, Euler-Lagrange equations as given by*

$$-[L^*]_{\boldsymbol{q}} = \mathrm{D}_t(\partial_{\dot{\boldsymbol{q}}} L^*) - \partial_{\boldsymbol{q}} L^* = \mathbf{0} \tag{3.38}$$

can be established. Eq. (3.38) *represents necessary conditions to the functions* $\boldsymbol{q}(t)$ *and* $\boldsymbol{\lambda}(t)$ *extremizing the augmented action*

$$S^* = \int_{t_0}^{t_1} L^*(t, \boldsymbol{q}, \mathrm{D}_t \boldsymbol{q}) \, \mathrm{d}t \, .$$

This is equivalent to (3.37) *along with* (3.35) *(cf. [32], [71] and [68]).*

Example 3.13 (Constrained mechanical systems). *Emphasizing the relevance of Rem. 3.12, the variational problem with (finite) constraints will be pursued subsequently for the special case of functionals depending on two independent functions subjected to holonomic constraints. The functions* $x : \mathcal{T} \mapsto \mathbb{R}$ *and* $y : \mathcal{T} \mapsto \mathbb{R}$ *are searched for such that the action*

$$S = \int_{t_0}^{t_1} L(t, x, y, x', y') \, \mathrm{d}t \tag{3.39}$$

is extremized, by additionally complying holonomic constraints as given by

$$\boldsymbol{g}(t, x, y) = 0 \, . \tag{3.40}$$

Geometrically, the curves $\boldsymbol{q}(t) = \begin{bmatrix} x(t) & y(t) \end{bmatrix}$*, lying on the constraint manifold* $C = \{\boldsymbol{q}(t) : \boldsymbol{g}(\boldsymbol{q}(t)) = \mathbf{0}\}$ *and extremizing the action* (3.39) *are searched for. Throughout this example the standard notation* $f'(t) = \mathrm{D}_t f(t)$ *is used to denote the derivative of a function* $f(t)$ *with respect to time* t*. Assuming non-singular constraint equations, i.e.* $\partial_x g \neq 0$ *and* $\partial_y g \neq 0$*, Eq.* (3.40) *might be resolved as*

$$y = k(t, x) \tag{3.41}$$

Consequently, the Lagrangian of the problem at hand reads as

$$L(t, x, y, x', y') = L(t, x, k, x', \partial_t k + (\partial_x k) x') \, . \tag{3.42}$$

In Eq. (3.42) *the time derivative of* (3.41) *as given by*

$$y' = \mathrm{D}_t k(t, x) = \partial_t k + (\partial_x k) x'$$

is used. In accordance to classical calculus of variations, a function $x(t)$ *that extremizes the action*

$$S = \int_{t_0}^{t_1} L(t, x, k, x', \partial_t k + (\partial_x k) x') \, \mathrm{d}t \, . \tag{3.43}$$

is searched for. Following Rem. 3.12, the first variation of the action (3.43) *is obtained as*

$$\begin{aligned}\delta S &= \mathrm{D}_\alpha \int_{t_0}^{t_1} L(t, x_\alpha, k, x'_\alpha, \partial_t k + (\partial_{x_\alpha} k)x'_\alpha)\ \mathrm{d}t \bigg|_{\alpha=0} \\ &= \int_{t_0}^{t_1} \eta\partial_x L + (\partial_k L)(\partial_x k)\eta + \eta'\partial_{x'} L + \partial_{k'} L \left(\partial^2_{tx} k + x'\partial^2_{xx} k + \eta'\partial_x k\right)\ \mathrm{d}t \\ &= \int_{t_0}^{t_1} (\partial_{x'} L + (\partial_{k'} L)(\partial_x k))\,\eta'\ \mathrm{d}t \\ &\quad + \int_{t_0}^{t_1} \left(\partial_x L + (\partial_k L)(\partial_x k) + \partial_{(k'} L \left(\partial^2_{tx} k + x'\partial^2_{xx} k\right)\right)\eta\ \mathrm{d}t = 0\ .\end{aligned} \tag{3.44}$$

Integrating the first integral of Eq. (3.44) *by parts gives rise to*

$$\mathrm{D}_t\left(\partial_{x'} L + (\partial_{y'} L)(\partial_x k)\right) - \left[\partial_x L + (\partial_y L)\ (\partial_x k) + \partial_{y'} L \left(\partial^2_{tx} k + x'\partial^2_{xx} k\right)\right]\ . \tag{3.45}$$

Applying the product rule to the term $\mathrm{D}_t\left((\partial_{y'} L)(\partial_x k)\right)$ *and observing that*

$$\mathrm{D}_t(\partial_x k) = \partial^2_{tx} k + x'\partial^2_{xx} k\ ,$$

Eq. (3.45) *can be reformulated as*

$$(\mathrm{D}_t(\partial_{x'} L) - \partial_x L) + \left(\mathrm{D}_t\left(\partial_{y'} L\right) - \partial_y L\right)\partial_x k = 0\ . \tag{3.46}$$

Since $g(t, x, k) = 0$ *need to be fulfilled at any time, the following applies*

$$\mathrm{D}_x g = \partial_x g + (\partial_y g)(\partial_x k) = 0\ . \tag{3.47}$$

A comparison of (3.46) *and* (3.47) *inevitably yields*

$$(\mathrm{D}_t(\partial_{x'} L) - \partial_x L) : \left(\mathrm{D}_t\left(\partial_{y'} L\right) - \partial_y L\right) = \partial_x g : \partial_y g\ . \tag{3.48}$$

Since the Jacobian of the constraint is assumed to be non-singular at any time, i.e.

$$\partial_x g(x, y, t) \neq 0 \quad \text{and} \quad \partial_y g(x, y, t) \neq 0\ \forall t \in \mathcal{T}\ ,$$

there needs to exist a proportionality factor $\lambda : \mathcal{T} \mapsto \mathbb{R}$ *such that the necessary conditions*

$$-[L]_x = \mathrm{D}_t(\partial_{x'} L) - \partial_x L = \lambda\partial_x g \qquad \text{and} \qquad -[L]_y = \mathrm{D}_t\left(\partial_{\dot{y}} L\right) - \partial_y L = \lambda\partial_y g \tag{3.49}$$

to the solutions $x(t)$ *and* $y(t)$ *apply. Eq.* (3.49) *along with the constraint equation* (3.40) *are also referred to as Lagrange's equations of motion of first kind. They represent a system of differential-algebraic equations, governing the motion of constrained mechanical systems.*

These equations of motion are equivalent to the Euler-Lagrange equations

$$\begin{aligned}&-[L^*]_x = \mathrm{D}_t(\partial_{x'} L^*) - \partial_x L^* = 0,\\ &-[L^*]_y = \mathrm{D}_t(\partial_{y'} L^*) - \partial_y L^* = 0 \quad \text{and}\\ &-[L^*]_\lambda = -\partial_\lambda L^* = -g(t, x, y) = 0,\end{aligned}$$

representing necessary conditions to the functions $x(t)$ and $y(t)$ as well as to the Lagrange multiplier $\lambda(t)$ extremizing the action

$$S^* = \int_{t_0}^{t_1} L^*(t, x, y, x', y') \, \mathrm{d}t \tag{3.50}$$

for the augmented Lagrangian $L^* = L - \lambda g$ For more details on variational problems subjected to (finite) constraints, see [32, pp.143-220], [71] and [68].

3.2. Inverse dynamics problem

While the classical direct dynamics problem aims to find the overall motion $\boldsymbol{q}(t)$ of a (constrained) mechanical system excited by a given actuation $\boldsymbol{f}(t)$, the inverse dynamics problem seeks for an actuation $\boldsymbol{f}(t)$ such that the motion of the mechanical system can partly be specified.

In this context, inverse dynamics problems might be interpreted as mechanical systems subjected to rheonomic holonomic constraints. Neglecting any further geometric constraints, the motion of such systems is governed by differential-algebraic equations in form of

$$\begin{aligned} \boldsymbol{M}\,\mathrm{D}_t^2\boldsymbol{q} + \boldsymbol{F}(\mathrm{D}_t\boldsymbol{q}, \boldsymbol{q}, t) + \boldsymbol{B}_f^T\boldsymbol{f}(t) &= \boldsymbol{0}\,, \\ \boldsymbol{h}(t, \boldsymbol{q}) = \boldsymbol{H}\boldsymbol{q} - \boldsymbol{\gamma}(t) &= \boldsymbol{0}\,. \end{aligned} \tag{3.51}$$

In $(3.51)_2$, the matrix $\boldsymbol{H}$ is of boolean type. Its purpose lies in extracting the prescribed degrees of freedom from the nodal configuration vector $\boldsymbol{q} : \mathcal{T} \mapsto \mathbb{R}^k$. The actuating components $\boldsymbol{f} : \mathcal{T} \mapsto \mathbb{R}^m$ then take the role of Lagrange multipliers enforcing the imposed servo-constraint[5] conditions $(3.51)_2$. Note that $m < k$ is assumed, since we are focusing on underactuated systems.

In contrast to (ideal)-orthogonal contact-constraints that have been discussed in Sec. 3.1, servo-constraints in general do not have collocation property[6]. Geometrically, this means that the Lagrange multipliers enforcing the constraints are not orthogonal to the constraint manifold anymore (cf. [22, 87]).

We will demonstrate, that the spatial disjunction of the constraint and the constraint forces, i.e. the non-standard construction of the constraint realization, causes differential-algebraic equations, that are characterized by either high differentiation index' or the appearance of (unstable) internal dynamics.

[5] Servo-constraints are also frequently referred to as control- or program-constraints, (cf. [79], [64] and [22], respectively.)

[6] E.g. concerning the imposition of essential boundary conditions in the finite element methods, the constraints are commonly assumed to have collocation property [3].

Although, servo-constraints could be successfully applied to low-dimensional underactuated mechanical systems such as cranes [4, 26, 75, 98] and manipulators with passive joints [14, 18, 23], the treatment of higher dimensional systems is still an open issue.

In particular, we face the problem that both, (unstable) internal dynamics as well as high differentiation index' along with high demands on the smoothness of the prescribed trajectory are affected adversely by the spatial discretization. This and its impact on the solvability of the problem at hand will be elaborated subsequently.

Following [24], the geometrical properties of the constraint realization might be specified in terms of

$$n = \operatorname{rank}(\boldsymbol{H}\boldsymbol{M}^{-1}\boldsymbol{B}_f^T) \tag{3.52}$$

such that three distinct cases of the orientation of the actuation $\boldsymbol{B}_f^T\boldsymbol{f}(t)$ on the constraint manifold $C = \{\boldsymbol{q} : \boldsymbol{g}(\boldsymbol{q}, t) = \boldsymbol{0}\}$ can be identified.

For $n = m$, the constraint realization is referred to as *(non-ideal) orthogonal*. In consequence, this leads to *differentially non-flat* systems where (unstable) *internal dynamics* may arise. View Fig. 3.8 illustrating *non-ideal orthogonal* constraint realizations in contrast to *ideal orthogonal* constraint realizations, depicted in Fig. 3.7. (cf. [56] or [26]). The notion of differential flatness (cf. Def. 3.23 and references therein) goes back to [38] while the concept thereof can be traced back to [49].

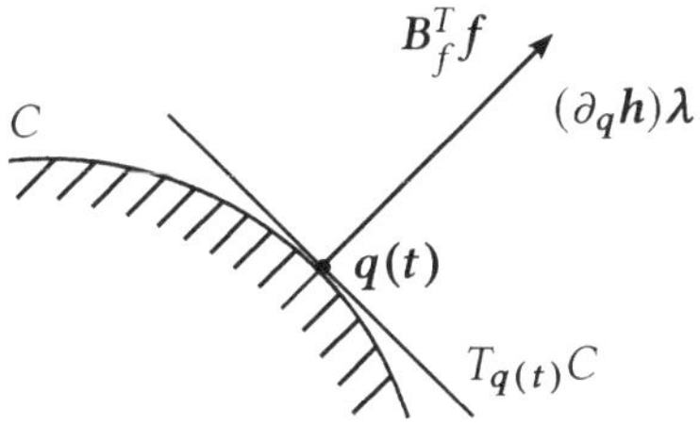

Figure 3.7.: Ideal orthogonal constraint representation.

For $n = 0$ or $0 < n < m$, the constraint realization is called *tangential* or *mixed orthogonal-tangential*, respectively. In the case of tangential constraint representations, the actuation $\boldsymbol{B}_f^T\boldsymbol{f}$ is tangential to the control constraint manifold $C = \{\boldsymbol{q} : \boldsymbol{g}(\boldsymbol{q}) = \boldsymbol{0}\}$, i.e. none of the constraint components can be actuated directly. In contrast to that, n constraint components can be actuated directly in the case of a mixed orthogonal-tangential representation. In both cases, the system at hand is possibly differentially flat without internal dynamics or non-flat with internal dynamics See Fig. 3.9 for a geometrical interpretation of the tangential constraint representation. In Tab. 3.1, the geometry of the constraint realization depending on Eq. (3.52) is summarized.

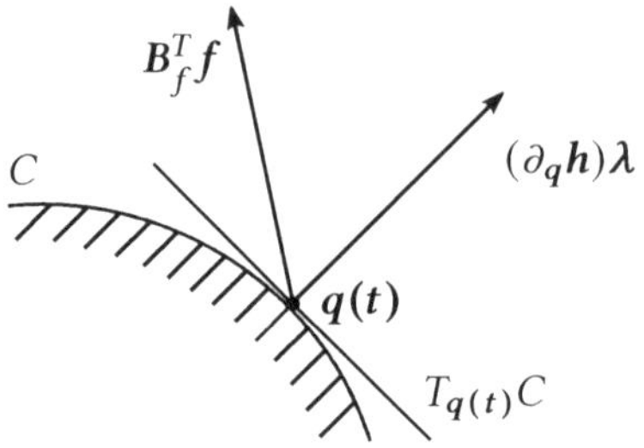

Figure 3.8.: Non-ideal orthogonal constraint representation.

In the case of differentially non-flat systems, (unstable) internal dynamics may occur, hindering numerical integration of the problem at hand. Therefore, it is inevitable to identify any (unstable) internal dynamics and carry out relevant analysis of the resulting *(non-) minimum phase* systems (cf. [17] and [85]). On the other hand, flat systems lead to differential-algebraic systems that are characterized by a high *differentiation index* ν_d (cf. Def. 3.7 and references therein). The concept of characterizing differential-algebraic systems of equations by its differentiation index ν_d is highly related to the concept of *relative degree* r (cf. Def. 3.14 and references therein).

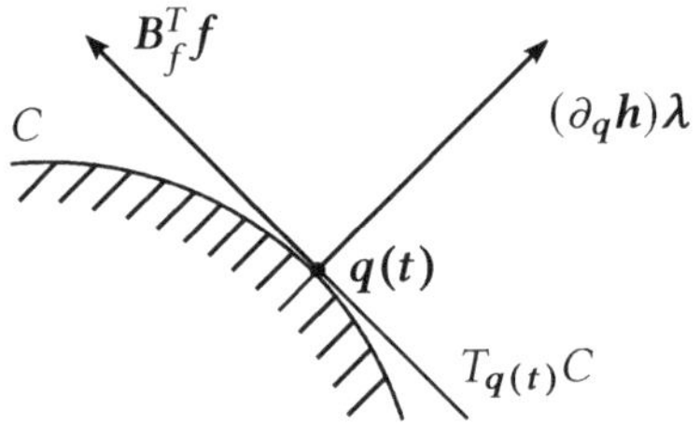

Figure 3.9.: Tangential constraint representation.

Since a numerically stable solution to the resulting differential-algebraic system of equations is depending significantly on the differentiation index, the differentiation index need to be reduced in order to achieve stable numerical solutions (cf. [4] and references therein).

(i) $n = m$: (non-ideal) orthogonal constraint realization, i.e. all m constraint components can be actuated

(ii) $0 < n < m$: mixed orthogonal-tangential constraint realization, i.e. only n constraint components can be actuated directly

(iii) $n = 0$: tangential constraint realization, i.e. none of the constraint components can be actuated directly

Table 3.1.: Geometry of the constraint realization depending on $n = \mathrm{rank}(HM^{-1}B_f^T)$

Subsequent examples will demonstrate that both internal dynamics and high differentiation index are a direct consequence of the discretization process. This obviously restricts the applicability of the classical semi-discretization approaches.

Definition 3.14 (Relative degree). *Consider a single input - single output system as given by*

$$\begin{aligned} \mathrm{D}_t \boldsymbol{z}(t) &= \boldsymbol{\mathcal{F}} + \boldsymbol{\mathcal{B}} u(t) \ , \\ y(t) &= \boldsymbol{\mathcal{H}}(\boldsymbol{z}) \ . \end{aligned} \tag{3.53}$$

System (3.53) *is said to have relative degree r at $\boldsymbol{z}_0$ if*

$$L_{\mathcal{B}} L_{\mathcal{F}} \boldsymbol{\mathcal{H}}(\boldsymbol{z}) = 0, \quad L_{\mathcal{B}} L_{\mathcal{F}}^2 \boldsymbol{\mathcal{H}}(\boldsymbol{z}) = 0, \quad \cdots \quad L_{\mathcal{B}} L_{\mathcal{F}}^i \boldsymbol{\mathcal{H}}(\boldsymbol{z}) = 0 \ \forall i < r-1 \tag{3.54}$$

along with a non-singular $L_{\boldsymbol{\mathcal{B}}} L_{\boldsymbol{\mathcal{F}}}^{r-1} \boldsymbol{\mathcal{H}}(\boldsymbol{z}) \neq 0$ applies. In Eq. (3.54),

$$L_{\mathcal{F}} \boldsymbol{\mathcal{H}}(\boldsymbol{z}) = \left. \frac{d}{d\alpha} \boldsymbol{\mathcal{H}}(\boldsymbol{z}) \right|_{\alpha=0}$$

denotes the Lie-derivative of $\boldsymbol{\mathcal{H}}$ along $\boldsymbol{\mathcal{F}}$ (cf. e.g. [56, Sec.1.2.]). The relative degree r of (3.53) *can be interpreted as the number of derivatives of the output $y(t)$ that has to be taken such that the input $\boldsymbol{\mathcal{F}}(t)$ appears explicitly (cf. [56, Sec.4.1. p.139]). With* (3.54)*, a coordinate transformation can be established such that system* (3.53) *can be transformed into a Byrnes-Isidori input-output normal form (cf. [86, Sec.3.1.]).*

Example 3.15 (Spring-mass system mounted on a moveable carriage). *The following example aims to underline the importance of the geometry of the constraint realizations. Therefore, we aim to analyze the following control problem: On a horizontally moveable carriage with mass $m_1 = 1$, a harmonic oscillator with mass $m_2 = 1$ and stiffness $k = 1$, rotated relatively to the carriage by an angle φ, is attached. Furthermore, the external load $F : \mathcal{T} \mapsto \mathbb{R}$ is assumed to act horizontally on the carriage. In Fig. 3.10, an illustration of the model is depicted.*

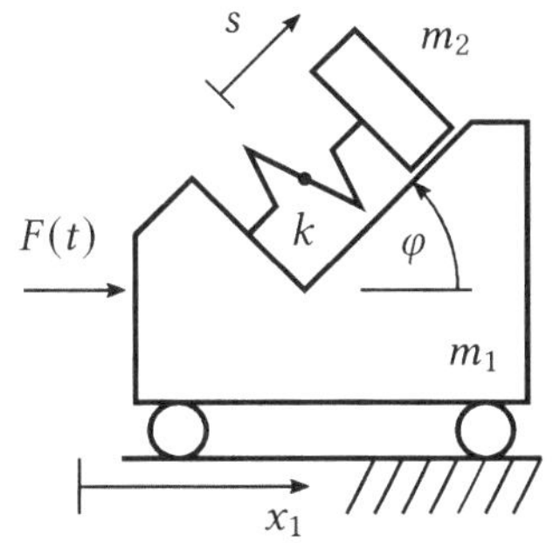

Figure 3.10.: Illustration of a spring-mass system mounted on a moveable carriage.

Consulting Rem. 3.12 the Lagrangian

$$L = \frac{v_1^2}{2} + \frac{v_2^2}{2} - \frac{s^2}{2} \tag{3.55}$$

gives rise to the Euler-Lagrange equations given by

$$2\,\mathrm{D}_t^2 x_1 + \cos(\varphi)\,\mathrm{D}_t^2 s - F = 0 \quad \textit{and} \quad \cos(\varphi)\,\mathrm{D}_t^2 x_1 + \mathrm{D}_t^2 s + s = 0\;. \tag{3.56}$$

In Eq. (3.55), *the velocities of the carriage and the mounted weight have been introduced as*

$$v_1 = \begin{bmatrix} \mathrm{D}_t x_1 \\ 0 \end{bmatrix} \quad \textit{and} \quad v_2 = \begin{bmatrix} \mathrm{D}_t x_1 + \cos(\varphi)\,\mathrm{D}_t s \\ \sin(\varphi)\,\mathrm{D}_t s \end{bmatrix}, \tag{3.57}$$

respectively. Regarding the inverse dynamics problem, the actuating force $F(t)$ is searched for, such that the horizontal motion x_2 of mass m_2 follows a prescribed trajectory as given by

$$h(x, s, t) = x_2 - \gamma(t) = x_1 + s\cos(\varphi) - \gamma(t) = 0\;.$$

Analyzing the geometry of the constraint realization, three distinct cases can be observed:

(i) $\varphi = 0$: *Ideal-orthogonal constraint realization - stable internal dynamics occur*

(ii) $0 < \varphi < \frac{\pi}{2}$: *Non-ideal orthogonal constraint realization*

(iii) $\varphi = \frac{\pi}{2}$: *Tangential constraint realization - no internal dynamics occur*

Further analysis regarding this control problem can be found in [23] and [87].

Example 3.16 (Linear elastic bar - inverse dynamics problem)**.** *Revisiting the linear-elastic bar, introduced in Ex. 2.5 and Ex. 3.1 for the pure Neumann problem, the semi-discrete equations of motion* (3.4) *in conjunction with servo-constraints boil down to a system of differential-algebraic equations given by*

$$\begin{aligned} M_{ij}\,\mathrm{D}_t^2 u_j + K_{ij} u_j &= -\delta_{i1} f\,, \\ u_{p+1} &= \gamma\;. \end{aligned} \tag{3.58}$$

Subsequently, a consistent approximation is compared to a lumped approximation of the inertia terms with regard to its impact on the properties of the constraint realization.

(i) Lumped mass matrix. Consider a mass matrix with entries $M_{ij} = M_i \delta_{ij}$. Regarding the last two nodes p and $p+1$ associated with the p-th element, the differential-algebraic system of equations (3.58) *yields*

$$\begin{aligned} M_p\,\mathrm{D}_t^2 u_p + K_{p,p} u_p + K_{p,p+1}\gamma &= 0 \\ M_{p+1}\,\mathrm{D}_t^2 \gamma + K_{p+1,p} u_p + K_{p+1,p+1}\gamma &= 0. \end{aligned} \tag{3.59}$$

Here, the summation convention does not apply. Since $\gamma(t)$ is prescribed, it can be deduced from the last equation that the displacement of node p is given by

$$u_p = -K_{p+1,p}^{-1}\left(M_{p+1}\,\mathrm{D}_t^2\gamma + K_{p+1,p+1}\gamma\right) \equiv u_p\left(\mathrm{D}_t^2\gamma, \gamma\right)\;. \tag{3.60}$$

Consequently,

$$D_t^2 u_p = a_p \left(D_t^4 \gamma, D_t^2 \gamma\right) \tag{3.61}$$

is obtained where the function a_p follows directly from differentiating (3.60) *twice with respect to time. Repeating this procedure for the elements $p-1,\dots,p-n$ reveals that*

$$u_{p-n} = u_{p-n}\left(D_t^{2n+2}\gamma, D_t^{2n}\gamma, \dots, \gamma\right), \quad D_t^2 u_{p-n} = a_{p-n}\left(D_t^{2n+4}\gamma, D_t^{2n+2}\gamma, \dots, D_t^2\gamma\right)$$

Setting $n = p-1$ leads to

$$u_1 = u_1\left(D_t^{2p}\gamma, D_t^{2p-2}\gamma, \dots, \gamma\right), \quad D_t^2 u_1 = a_1\left(D_t^{2p+2}\gamma, D_t^{2p}\gamma, \dots, D_t^2\gamma\right) \tag{3.62}$$

so that for $i=1$, $(3.58)_1$ *yields*

$$f = f\left(D_t^{2p+2}\gamma, D_t^{2p}\gamma, \dots, D_t^2\gamma, \gamma\right) . \tag{3.63}$$

Differentiating Eq. (3.63) *once again with respect to time, reveals that the time derivative of the algebraic variable f pertaining to the differential-algebraic system of equations* (3.58) *can be obtained after $2p+3$ time derivatives of the algebraic constraint* $(3.58)_2$*. This implies that the system* (3.58) *has differentiation index $\nu_d = 2p+3$ (cf. [13, 63]). Accordingly, the index increases with the number p of finite elements used for the space discretization of the bar. Since all of the unknowns of system* (3.58) *can be expressed in terms of γ and derivatives thereof, the present problem exhibits the property of differential flatness ([67, 90]). In particular, the nodal displacement u_{p+1} plays the role of a flat output. Analogous results have been obtained for the articulated mass point systems dealt with in [22, 26, 74]. Note that the simple result for the flat output in Ex. 3.16 hinges on the mass matrix being diagonal. The situation gets more intricate in case of a consistent mass matrix or higher-order finite elements.*

Special case $n = 1$. *Using only one finite element, the problem boils down to the system of ordinary differential equations*

$$\begin{aligned} D_t^2 u_1 + u_1 - u_2 &= -f , \\ D_t^2 u_2 - u_1 + u_2 &= 0 . \end{aligned}$$

subjected to a rheonomic holonomic constraint given by

$$g(u_2) = u_2 - \gamma(t) = 0 .$$

Essentially, this differential-algebraic system of equations governs the inverse dynamics of a vibrating (double-) oscillator. By introducing

$$\boldsymbol{M} = \begin{bmatrix} 1 & 0 \\ 0 & 1 \end{bmatrix}, \quad \boldsymbol{F} = \begin{bmatrix} 1 & -1 \\ -1 & 1 \end{bmatrix}\begin{bmatrix} u_1 \\ u_2 \end{bmatrix}, \quad \boldsymbol{B}_f = \begin{bmatrix} 1 \\ 0 \end{bmatrix} \quad \textit{and} \quad \boldsymbol{H} = \begin{bmatrix} 0 \\ 1 \end{bmatrix}$$

as well as the actuating force f acting as Lagrange multiplier, the resulting system aligns with system (3.51). *Due to the non-collocation of the constraint force and the imposed constraint, the constraint realization is referred to as tangential. In the case of this two-dimensional system, the constraint realization can be visualized in Fig. 3.11.*

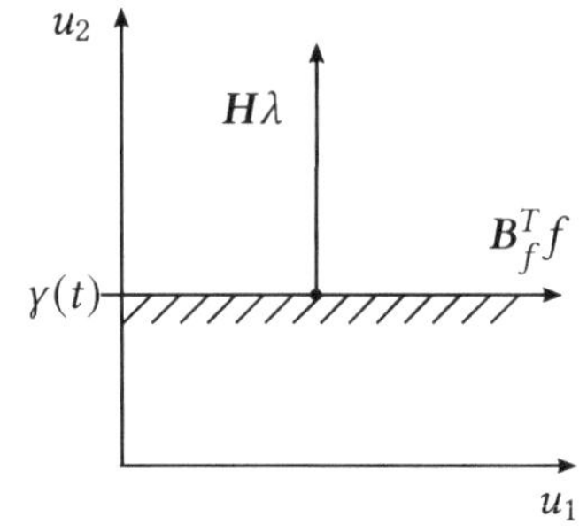

Figure 3.11.: Tangential realization of the servo-constraint.

(ii) *Consistent mass matrix. For the more general case of a consistent mass matrix (cf. Ex. 3.1) and confining to piecewise linear Lagrangian shape functions $L_j(s)$ it can be observed that due to $\boldsymbol{H} = \begin{bmatrix} 0 & 0 & 0 & \cdots & 0 & 1 \end{bmatrix}$ and $\boldsymbol{B}_f = \begin{bmatrix} 1 & 0 & 0 & \cdots & 0 & 0 \end{bmatrix}$ the norm of $\boldsymbol{P} = \boldsymbol{H}\boldsymbol{M}^{-1}\boldsymbol{B}_f^T$ depends on the number of elements p. Apparently, $\boldsymbol{P}$ becomes singular as p tends towards ∞. Hence, by increasing the number of elements, the constraint realization tends to become tangential. (cf. Fig. 3.12)*

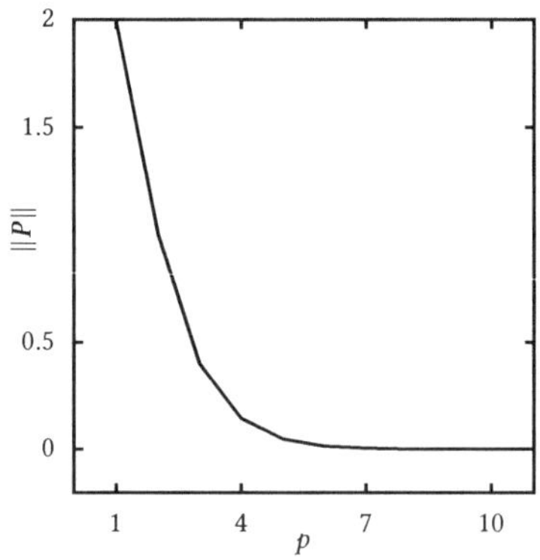

Figure 3.12.: Norm of $\boldsymbol{P} = \boldsymbol{H}\boldsymbol{M}^{-1}\boldsymbol{B}_f^T$ depending on the number of elements p.

Example 3.17 (Overhead crane - inverse dynamics problem). *Revisiting the model of the planar overhead crane introduced in Ex. 3.11, the corresponding inverse dynamics problem* *aims to find an actuation* $\boldsymbol{f} = \begin{bmatrix} F(t) & M(t) \end{bmatrix}^T$ *such that the attached load follows a prescribed trajectory*

$$x(t) = \gamma_x(t) ,$$
$$z(t) = \gamma_z(t) .$$

Similar to the formulation of the direct problem, the inverse dynamics can be established either in redundant or in generalized coordinates.

(i) *Redundant coordinates. Using redundant coordinates* $\boldsymbol{q} = \begin{bmatrix} s & l & x & z \end{bmatrix}^T$, *the control problem can be formulated in terms of a differential-algebraic system of equations as given by*

$$\begin{aligned} &\mathrm{D}_t^2 \boldsymbol{q} + \boldsymbol{F} - \boldsymbol{G}^T \lambda - \boldsymbol{B}_f^T \boldsymbol{f} = \boldsymbol{0} , \\ &g(\boldsymbol{q}) = \frac{1}{2}((x-s)^2 + z^2 - l^2) = 0 , \\ &\boldsymbol{h}(\boldsymbol{q}) = \boldsymbol{0} \end{aligned} \tag{3.64}$$

where $\boldsymbol{F} = \begin{bmatrix} 0 & 0 & 0 & 1 \end{bmatrix}^T$ *denotes the external applied excitation, excluding the actuation* $\boldsymbol{f}$. *Additionally to the geometric, i.e. ideal-orthogonal constraint* $g(\boldsymbol{q})$, *that is enforced by the constraint forces* $\boldsymbol{G}\lambda = (\partial_{\boldsymbol{q}} g)\lambda$ *(cf. Ex. 3.11), the system at hand is subjected to servo-constraints*

$$\boldsymbol{h}(\boldsymbol{q}) = \begin{bmatrix} x(t) - \gamma_x(t) \\ z(t) - \gamma_z(t) \end{bmatrix} , \tag{3.65}$$

that are enforced by the external actuation $\boldsymbol{B}_f \boldsymbol{f}$. *Due to the non-collocation of*

$$\boldsymbol{B}_f = \begin{bmatrix} 1 & 0 & 0 & 0 \\ 0 & 1 & 0 & 0 \end{bmatrix}$$

and the Jacobian of the servo-constraint

$$\partial_{\boldsymbol{q}} \boldsymbol{h}(\boldsymbol{q}) = \boldsymbol{H} = \begin{bmatrix} 0 & 0 & 1 & 0 \\ 0 & 0 & 0 & 1 \end{bmatrix} \tag{3.66}$$

it applies, that $n = \operatorname{rank}(\boldsymbol{H}\boldsymbol{M}^{-1}\boldsymbol{B}_f^T) = 0$. *Hence the realization of the servo-constraint can be identified to be tangential (cf. Fig. 3.9). As a consequence, the differentiation index of the underlying differential-algebraic system of equations can be determined to be* $\nu_d = 5$.

(ii) *Generalized coordinates. In the case of using generalized coordinates* $\boldsymbol{q} = \begin{bmatrix} s & l & \varphi \end{bmatrix}^T$, *the inverse dynamics problem is given by the differential-algebraic system of equations*

$$\begin{aligned} &\boldsymbol{M}\,\mathrm{D}_t^2 \boldsymbol{q} = \boldsymbol{F} - \boldsymbol{B}_f^T \boldsymbol{f} , \\ &\boldsymbol{h}(\boldsymbol{q}) = \boldsymbol{0} \end{aligned} \tag{3.67}$$

for

$$h(q) = \begin{bmatrix} x - \gamma_x \\ z - \gamma_z \end{bmatrix} = \begin{bmatrix} s + l\sin(\varphi) - \gamma_x \\ l\cos(\varphi) - \gamma_z \end{bmatrix} = \mathbf{0} \tag{3.68}$$

and

$$\boldsymbol{B}_f = \begin{bmatrix} 1 & 0 & 0 \\ 0 & 1 & 0 \end{bmatrix}. \tag{3.69}$$

Again, the non-collocation of $\boldsymbol{B}_f$ *and*

$$\partial_q \boldsymbol{h}(\boldsymbol{q}) = \boldsymbol{H} = \begin{bmatrix} 1 & \sin(\varphi) & l\cos(\varphi) \\ 0 & \cos(\varphi) & -l\sin(\varphi) \end{bmatrix}$$

leads to $\mathrm{rank}(\boldsymbol{H}\boldsymbol{M}^{-1}\boldsymbol{B}_f^T) = 1$ *and consequently to a mixed orthogonal-tangential constraint realization.*

Remark 3.18 (Index reduction by minimal extension). *By augmenting the differential-algebraic system at hand by feasible index 1 systems the differentiation index can be reduced. Since the resulting system is singular so-called dummy variables, i.e. additional unknown functions, need to be introduced. This concept of index reduction by minimal extension goes back to [73] and is motivated by the GGL stabilization [42].*

Example 3.19. *Consider a Hessenberg system with index* $\nu_d = 3$*, taken from [73], as given by*

$$\begin{aligned} \mathrm{D}_t u(t) &= v(t)\,, \\ \mathrm{D}_t v(t) &= f(t)\,, \\ g(u,t) = u(t) - \gamma(t) &= 0\,. \end{aligned} \tag{3.70}$$

This system might be augmented by successive derivatives of $(3.70)_1$ *and* $(3.70)_3$*, i.e. by a differential chain in index 1 form given by*

$$\begin{aligned} \mathrm{D}_t u(t) &= \mathrm{D}_t \gamma(t)\,, \\ \mathrm{D}_t^2 u(t) &= \mathrm{D}_t^2 \gamma(t)\,, \\ \mathrm{D}_t v(t) &= \mathrm{D}_t^2 u(t)\,. \end{aligned} \tag{3.71}$$

System. (3.70) *along with system* (3.71) *form an overdetermined system for the unknown functions* $u(t)$*,* $v(t)$ *and* $f(t)$*. Therefore, dummy variables* $\xi_1 \equiv \mathrm{D}_t u(t)$*,* $\xi_2 \equiv \mathrm{D}_t^2 u(t)$ *and* $\xi_3 \equiv \mathrm{D}_t v(t)$ *need to be introduced in order to be able to solve the resulting system of purely algebraic equations (*$\nu_d = 1$*) uniquely.*

Example 3.20 (Bead on parabola). *The motion of a bead sliding on a parabola is governed by a system of differential-algebraic equations given by*

$$\begin{aligned} \mathrm{D}_t u_1 - v_1 &= 0\,, \\ \mathrm{D}_t u_2 - v_2 &= 0\,, \\ \mathrm{D}_t v_1 - 2\lambda u_1 &= 0\,, \\ \mathrm{D}_t v_2 - 1 + \lambda &= 0\,, \\ g(u_1, u_2) = u_2 - u_1^2 &= 0\,. \end{aligned} \tag{3.72}$$

System (3.72) is characterized by a differentiation index $\nu_d = 3$. Augmenting system (3.72) by the first and second time derivative of the constraint equation $(3.72)_5$ with respect to the time variable, i.e.

$$\mathrm{D}_t g(u_1, u_2) = v_2 - 2v_1 u_1 = 0$$

and

$$\mathrm{D}_t^2 g(u_1, u_2) = \mathrm{D}_t v_2 - 2\,\mathrm{D}_t v_1 u_1 - 2\,\mathrm{D}_t v_1^2 = -(1+\lambda) - 4\lambda u_1^2 - 2v_1 = 0,$$

respectively, it can be transformed into a fully algebraic system of equations, i.e. $\nu_d = 1$, by introducing the new variables $\xi_1 \equiv \mathrm{D}_t u_2$ and $\xi_2 \equiv \mathrm{D}_t v_2$ (cf. [63, Sec. 6.4]).

Example 3.21 (Mass-spring system - index reduction by minimal extension). *Revisiting Ex. 3.16 and considering the special case of using one finite element for the spatial approximation of the solution, yields*

$$\begin{aligned} \mathrm{D}_t^2 u_1(t) + u_1(t) - u_2(t) - f(t) &= 0\,, \\ \mathrm{D}_t^2 u_2(t) - u_1(t) + u_2(t) &= 0\,, \\ g(u_2) = u_2 - \gamma(t) &= 0\,. \end{aligned} \tag{3.73}$$

The system of differential-algebraic equations (3.73) features a differentiation index $\nu_d = 5$. Augmenting system (3.73) by $\mathrm{D}_t^2 u_2(t) = \mathrm{D}_t^2 \gamma(t) \equiv \xi_1(t)$ yields

$$\begin{aligned} \mathrm{D}_t^2 u_1(t) + u_1(t) - u_2(t) - f(t) &= 0\,, \\ \xi_1(t) - u_1(t) + u_2(t) &= 0\,, \\ g(u_2) = u_2 - \gamma(t) &= 0\,, \\ \xi_1(t) &= \mathrm{D}_t^2 \gamma(t). \end{aligned} \tag{3.74}$$

A Hessenberg form with differentiation index $\nu_d = 3$ can be identified by introducing the velocity $v_1 \equiv \mathrm{D}_t u_1$ (cf. Def. 3.6). Augmenting system (3.74) by

$$\mathrm{D}_t^2 u_1(t) = \mathrm{D}_t^2 \gamma(t) + \gamma(t) \equiv \xi_2(t) \tag{3.75}$$

a fully algebraic system of equations featuring $\nu_d = 1$ as given by

$$\begin{aligned} \xi_2(t) + u_1(t) - u_2(t) - f(t) &= 0\,, \\ \xi_1(t) - u_1(t) + u_2(t) &= 0\,, \\ g(u_2) = u_2 - \gamma(t) &= 0\,, \\ \xi_1(t) &= \mathrm{D}_t^2 \gamma(t)\,, \\ \xi_2(t) &= \mathrm{D}_t^2 \gamma(t) + \gamma(t) \end{aligned} \tag{3.76}$$

can be established. By using the constraint equation $(3.73)_3$, the prescribed function u_2 serves as a parametrization, to transform the differential part of system (3.73) into a purely algebraic system of equations and finally solve the problem at hand without actually integrating. In this context, the parameterization u_2 serves as a differentially flat output algebraizing the differential-algebraic system of equations (cf. Def. 3.23 as well as accompanying Ex. 3.24, Ex. 3.25 and Ex. 3.26 for more details on differential flatness). This indicates a conceivable minimal extension of the equations, aiming to reduce the differentiation index.

Example 3.22 (Overhead crane - index reduction by minimal extension). *Revisiting the overhead crane model introduced in Ex. 3.11 and Ex. 3.17 in the context of the direct and inverse dynamics problem, respectively. In case of choosing redundant coordinates* $\boldsymbol{q} = \begin{bmatrix} s & l & x & z \end{bmatrix}^T$, *the actuation* $\boldsymbol{f}(t)$ *satisfying the system of differential-algebraic equations given by*

$$\begin{aligned} \mathrm{D}_t^2 \boldsymbol{q} - \partial_{\boldsymbol{q}} g(\boldsymbol{q})\lambda - \boldsymbol{B}_f \boldsymbol{f} &= 0 \\ g(\boldsymbol{q}) &= 0 \\ \boldsymbol{h}(\boldsymbol{q}) &= 0 \end{aligned} \tag{3.77}$$

is searched for. Augmenting system (3.77) *by the second time derivative of the servo-constraint given by*

$$\mathrm{D}_t^2 \boldsymbol{h}(\boldsymbol{q}) = 0$$

and introducing the new variables $\xi_1 \equiv \mathrm{D}_t^2 x$ *and* $\xi_1 \equiv \mathrm{D}_t^2 z$, *yields a system with differentiation index* $\nu_d = 3$. *An algebraic relation for the Lagrange multiplier* λ *is then given by*

$$\lambda = \frac{1}{\gamma_2}(1 - \xi_2) \, . \tag{3.78}$$

Definition 3.23 (Differential flatness, cf. [49]). *A differential equation of the form*

$$F(t, u(t), \mathrm{D}_t u, \cdots, \mathrm{D}_t^n u, f(t), \mathrm{D}_t f, \cdots, \mathrm{D}_t^n f) = 0 \tag{3.79}$$

is called differentially flat, if there exists an arbitrary function $\xi \in \mathbb{R} \mapsto w(\xi) \in \mathbb{R}$ *with derivatives* $\mathrm{D}_\xi w(\xi), \cdots, \mathrm{D}_\xi^n w(\xi)$ *such that a transformation of the form*

$$\begin{aligned} u &= a(\xi, w, \mathrm{D}_\xi w, \mathrm{D}_\xi^2 w, \cdots) \\ f &= b(\xi, w, \mathrm{D}_\xi w, \mathrm{D}_\xi^2 w, \cdots) \\ t &= c(\xi, w, \mathrm{D}_\xi w, \mathrm{D}_\xi^2 w, \cdots) \end{aligned} \tag{3.80}$$

satisfies the differential equation (3.79) *by taking into account*

$$\begin{aligned} \mathrm{D}_t u(t) &= \mathrm{D}_\xi a \, \mathrm{D}_c \xi = \frac{\mathrm{D}_\xi a}{\mathrm{D}_\xi c}, & \mathrm{D}_t^2 u(t) &= \frac{1}{(\mathrm{D}_\xi c)^2}(\mathrm{D}_\xi^2 a \, \mathrm{D}_\xi c - \mathrm{D}_\xi a \, \mathrm{D}_\xi^2 c), & \cdots \\ \mathrm{D}_t f(t) &= \mathrm{D}_\xi b \, \mathrm{D}_c \xi = \frac{\mathrm{D}_\xi b}{\mathrm{D}_\xi c}, & \mathrm{D}_t^2 f(t) &= \frac{1}{(\mathrm{D}_\xi c)^2}(\mathrm{D}_\xi^2 b \, \mathrm{D}_\xi c - \mathrm{D}_\xi b \, \mathrm{D}_\xi^2 c), & \cdots \end{aligned}$$

along with

$$\begin{aligned} \mathrm{D}_\xi a(\xi, w, \mathrm{D}_\xi w, \mathrm{D}_\xi^2 w, \cdots) &= \partial_\xi a + \partial_w a \, \mathrm{D}_\xi w + \partial_{(\mathrm{D}_\xi w)} a \, \mathrm{D}_\xi^2 w + \cdots \\ \mathrm{D}_\xi b(\xi, w, \mathrm{D}_\xi w, \mathrm{D}_\xi^2 w, \cdots) &= \partial_\xi b + \partial_w b \, \mathrm{D}_\xi w + \partial_{(\mathrm{D}_\xi w)} b \, \mathrm{D}_\xi^2 w + \cdots \quad . \end{aligned}$$

Alternative definitions of differential flatness can be found e.g. in [67] and [38]. See also [90] for more background on linear systems.

Example 3.24 (Differential flatness). *Consider the first-order ordinary differential equation*

$$F(t, \mathrm{D}_t u, f) = \mathrm{D}_t u(t) - f(t) = 0 \,. \tag{3.81}$$

The transformation

$$\begin{aligned} u &= a(\xi, w, \mathrm{D}_\xi w, \mathrm{D}_\xi^2 w, \cdots) = w(\xi) \,, \\ f &= b(\xi, w, \mathrm{D}_\xi w, \mathrm{D}_\xi^2 w, \cdots) = \mathrm{D}_\xi w(\xi) \,, \\ t &= c(\xi, w, \mathrm{D}_\xi w, \mathrm{D}_\xi^2 w, \cdots) = \xi \end{aligned}$$

with

$$\mathrm{D}_t u(t) = \mathrm{D}_\xi w(\xi)$$

satisfies the differential equation (3.81).

Example 3.25 (Differential flatness). *Consider the second-order ordinary differential equation*

$$F(t, \mathrm{D}_t^2 u, \mathrm{D}_t u, f) = \mathrm{D}_t^2 u(t) + u(t) - f(t) = 0 \,. \tag{3.82}$$

The transformation

$$\begin{aligned} u &= a(\xi, w) = w(\xi) \\ f &= b(\xi, w, \mathrm{D}_\xi w, \mathrm{D}_\xi^2 w) = \mathrm{D}_\xi^2 w(\xi) + w(\xi) \\ t &= c(\xi) = \xi \end{aligned}$$

with

$$\mathrm{D}_t u(t) = \mathrm{D}_\xi w(\xi) \quad \textit{and} \quad \mathrm{D}_t^2 u(t) = \mathrm{D}_\xi^2 w(\xi)$$

satisfies the differential equation (3.82).

Example 3.26 (Differential flatness). *Consider the system of second-order ordinary differential equations*

$$\begin{aligned} \mathrm{D}_t^2 u_1 + u_1 - u_2 &= f \,, \\ \mathrm{D}_t^2 u_2 + u_2 - u_1 &= 0 \end{aligned} \tag{3.83}$$

governing the motion of a mass-spring system analyzed in Ex. 3.16 and Ex. 3.21. The transformation

$$\begin{aligned} u_1 &= a(\xi, w, \mathrm{D}_\xi^2 w) = \mathrm{D}_\xi^2 w(\xi) + w(\xi) \,, \\ u_2 &= b(\xi, w(\xi)) = w(\xi) \,, \\ t &= c(\xi) = \xi \end{aligned} \tag{3.84}$$

allows solving the differential equation without actually integrating it. Inserting the transformation (3.84) *satisfying* $(3.83)_2$ *into* $(3.83)_1$ *yields a relation for the function* $f : \mathcal{T} \mapsto R$ *given by*

$$f = d(\xi, \mathrm{D}_\xi^4 w(\xi), \mathrm{D}_\xi^2 w(\xi)) = \mathrm{D}_\xi^4 w(\xi) + 2\,\mathrm{D}_\xi^2 w(\xi) \,. \tag{3.85}$$

Remark 3.27 (A spectral Galerkin approach). *Alternatively, for interpolating the test and trial functions* (3.3) *in the weak formulation* (3.2), *global trigonometric basis functions of the form*

$$L_1(s) = 1, \quad L_2(s) = \frac{s^2}{L^2} - 2\frac{s}{L} + 1 \quad \text{and} \quad L_j(s) = \sin(\frac{j\pi}{L}) - (-1)^j \frac{j\pi}{L^2}(s^2 - sL)$$

for $j = 1, \cdots, p$ can be used.[7] *See Fig. 3.13 for an illustration of these basis functions. Subsequently it will be demonstated that indeed high differentiation index might be circumvented but eventually (unstable) internal dynamics occur.*

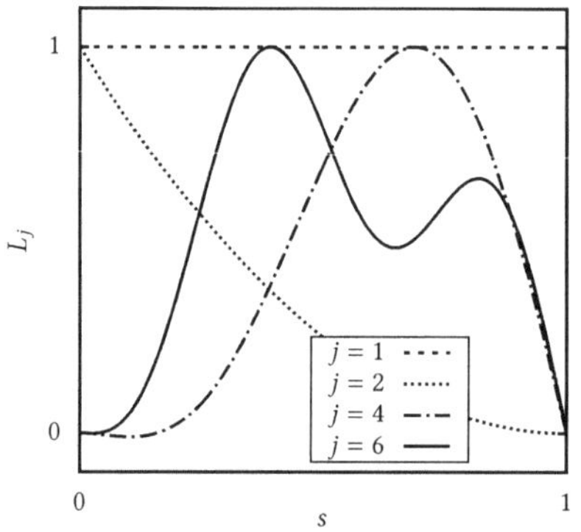

Figure 3.13.:
Global trigonomteric basis functions for $p = 5$

The resulting differential-algebraic system of equations can then be obtained as

$$\begin{aligned} M_{ij}\, \mathrm{D}_t^2 d_j + K_{ij} d_j - B_f^T f &= 0\,, \\ h = \gamma(t) - H_i d_i &= 0\,. \end{aligned} \tag{3.86}$$

In Eq. (3.86), *the two Boolean matrices B_f and H are defined as*

$$B_f = \begin{bmatrix} I & I & 0 & \cdots & 0 \end{bmatrix} \quad \text{and} \quad H = \begin{bmatrix} I & 0 & 0 & \cdots & 0 \end{bmatrix}, \tag{3.87}$$

respectively. Furthermore $K_{1j} = K_{j1} = 0$ applies. If $n = \operatorname{rank}(HM^{-1}B_f^T) = m$, where m is the number of the actuating components, the constraint realization is called (non-ideal) orthogonal (cf. Fig. 3.14 and Fig. 3.15 illustrating the difference of ideal- and non-ideal-orthogonal constraint realization). Orthogonal constraint realizations generally lead to differentially non-flat systems where (unstable) internal dynamics may arise[8] *(cf. [56] or [26]).*

[7] The idea of using this trigonometric ansatz functions has been suggested by R. Altmann after the talk [94].

[8] For $n = 0$ or $0 < n < m$, the constraint realization is called tangential or mixed orthogonal-tangential respectively. In both cases, the system at hand can be differentially flat (without internal dynamics) or non-flat (with internal dynamics).

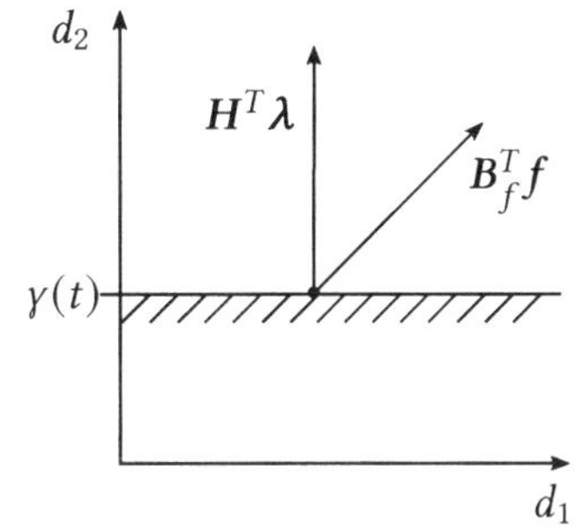

Figure 3.14.: Non-ideal- orthogonal constraint realization.

Due to the positive definiteness of the mass matrix $\boldsymbol{M}$ *for* (3.86) *it applies that* $n = m$. *Thus the control is non-ideal orthogonal to the constraint mainfold (tangential components exists, cf.* (3.88)*). Hence, the differentiation index* $\nu_d = 3$. *Therefore the output cannot be differentially flat and (unstable) internal dynamics may occur.*

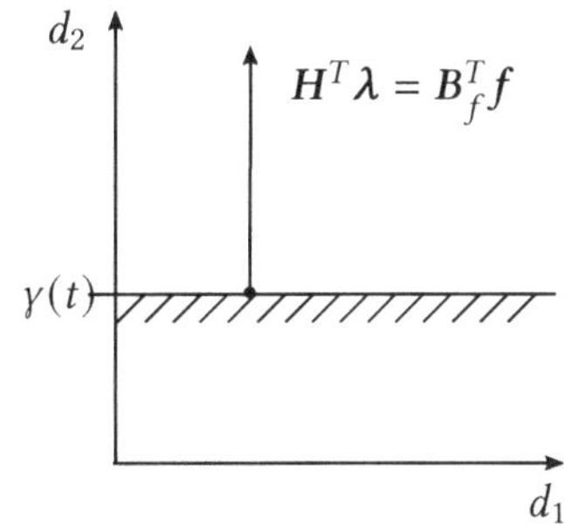

Figure 3.15.: Ideal- orthogonal constraint realization.

The internal dynamics of the system (3.86) *may be identified by taking the second time derivative of the algebraic constraint* $(3.86)_2$. *Inserting it into the differential part* $(3.86)_1$ *the actuation* $\boldsymbol{f}(t)$ *can be established as*

$$\boldsymbol{f} = \boldsymbol{M}_{11}\,\mathrm{D}_t^2\boldsymbol{\gamma} + \boldsymbol{M}_{12}\,\mathrm{D}_t^2\boldsymbol{d}_2 + \boldsymbol{M}_{13}\,\mathrm{D}_t^2\boldsymbol{d}_3 + \cdots \,. \tag{3.88}$$

A dependency of $\boldsymbol{f}(t)$ *on the unknown functions* $\boldsymbol{d}_i(t)$ *for* $i \in \{2, 3, ...\}$ *can be observed. By inserting the relation* (3.88) *into the remaining differential part of* (3.86), *the solution* $\boldsymbol{d}_i(t)$ *is governed by a system of ordinary differential equations given by*

$$\begin{bmatrix} \boldsymbol{M}_{22} - \boldsymbol{M}_{12} & \boldsymbol{M}_{23} - \boldsymbol{M}_{13} & \cdots \\ \boldsymbol{M}_{32} - \boldsymbol{M}_{12} & \boldsymbol{M}_{33} - \boldsymbol{M}_{13} & \cdots \\ \vdots & \vdots & \ddots \end{bmatrix} \begin{bmatrix} \mathrm{D}_t^2\boldsymbol{d}_2 \\ \mathrm{D}_t^2\boldsymbol{d}_3 \\ \vdots \end{bmatrix} + \begin{bmatrix} \boldsymbol{K}_{22} & \boldsymbol{K}_{23} & \cdots \\ \boldsymbol{K}_{32} & \boldsymbol{K}_{33} & \cdots \\ \vdots & \vdots & \ddots \end{bmatrix} \begin{bmatrix} \boldsymbol{d}_2 \\ \boldsymbol{d}_3 \\ \vdots \end{bmatrix} = \begin{bmatrix} (\boldsymbol{M}_{21} - \boldsymbol{M}_{11})\boldsymbol{\gamma}(t) \\ (\boldsymbol{M}_{31} - \boldsymbol{M}_{11})\boldsymbol{\gamma}(t) \\ \vdots \end{bmatrix} . \tag{3.89}$$

The system (3.89) *constitutes the internal dynamics of the control problem at hand (cf. [26]). Further investigations, regarding the stability of the internal dynamics* (3.89) *are mandatory.*

The existence of internal dynamics indicates that the present output (i.e. the displacement of the elastic rod at $s = L$) cannot be a flat output of the discrete system at hand. This observation leads to the question whether there does exist a flat output. That is, given the discrete system with actuating force at the left end, does there exist an output which turns the system differentially flat. This might be figured out by introducing $\boldsymbol{v}(t) \equiv \mathrm{D}_t \boldsymbol{d}(t)$. Then the equation of motion subjected to the servo-constraint can be transformed into a first-order system in state-space representation

$$\mathrm{D}_t \boldsymbol{x} = \boldsymbol{A}\boldsymbol{x} + \boldsymbol{b}\boldsymbol{f} \tag{3.90}$$

with

$$\boldsymbol{x} = \begin{bmatrix} \boldsymbol{d} \\ \boldsymbol{v} \end{bmatrix}, \qquad \boldsymbol{A} = \begin{bmatrix} \boldsymbol{I} & \boldsymbol{0} \\ \boldsymbol{0} & \boldsymbol{M} \end{bmatrix}^{-1} \begin{bmatrix} \boldsymbol{0} & \boldsymbol{I} \\ \boldsymbol{K} & \boldsymbol{0} \end{bmatrix} \qquad \text{and} \qquad \boldsymbol{b} = \begin{bmatrix} \boldsymbol{0} \\ \boldsymbol{B}_f^T \end{bmatrix}.$$

Following [90], a state-space coordinate transformation of the form $\boldsymbol{z} = \boldsymbol{T}\boldsymbol{x}$ can be carried out by using the inverse of Kalmans controllability matrix

$$\boldsymbol{T} = \begin{bmatrix} \boldsymbol{b} & \boldsymbol{A}\boldsymbol{b} & \boldsymbol{A}^2\boldsymbol{b} & \cdots \end{bmatrix}^{-1}. \tag{3.91}$$

This yields a system of ordinary differential equations in new coordinates as given by

$$\mathrm{D}_t \boldsymbol{z} = \boldsymbol{\Lambda}\boldsymbol{z} + \boldsymbol{T}\boldsymbol{b}\boldsymbol{f}. \tag{3.92}$$

For the transformed system, it is possible to find the differentially flat output corresponding to the actuation $\boldsymbol{f}$ (cf. Ex. 3.16 for $p = 2$ and Ex. 3.29 for $p = 3$).

The proposed global trigonometric discretization of the weak formulation leads to a differential-algebraic system characterized by a differentiation index $\nu_d = 3$. However, the actuating force $\boldsymbol{f}(t)$ cannot be identified as a differentially flat output corresponding to the partly specified motion $\boldsymbol{\gamma}(t)$. Consequently, internal dynamics that are eventually unstable occur. Further investigations should be pursued to elucidate the implications the internal dynamics may cause. Incidentally, a discretization of the weak formulation at hand by using Lagrangian shape functions assuming a consistent mass matrix also leads, at least for small p, to differential-algebraic equations with differentiation index $\nu_d = 3$ and internal dynamics (cf. Ex. 3.28).

Example 3.28 (Linear elastic rod - trigonometric discretization for $p = 2$). *Consider a linear elastic rod with length $L = 1$, axial stiffness $EA = 1$ and mass density $\rho A = 1$. By using $p = 2$ ansatz functions, the coefficients of the semi-discrete equations of motion can be identified as*

$$\boldsymbol{M} = \begin{bmatrix} 1 & 1/3 \\ Sym. & 1/5 \end{bmatrix}, \qquad \boldsymbol{K} = \begin{bmatrix} 0 & 0 \\ Sym. & 4/3 \end{bmatrix}, \qquad \boldsymbol{B}_f^T = \begin{bmatrix} 1 \\ 1 \end{bmatrix} \qquad \text{and} \qquad \boldsymbol{H}^T = \begin{bmatrix} 1 \\ 0 \end{bmatrix}.$$

Due to

$$n = rank(\boldsymbol{H}\boldsymbol{M}^{-1}\boldsymbol{B}_f^T) = 1$$

the actuator is (non-ideal) orthogonal to the constraint manifold (cf. Fig. 3.11). Differentiating the algebraic constraint twice and inserting it into the differential part of Eq. (3.86) *the actuating component is obtained as*

$$f(t) = \frac{1}{3}\,\mathrm{D}_t^2 d_2(t) + \mathrm{D}_t^2\gamma(t)\ . \tag{3.93}$$

Taking the time derivative of (3.93), *an ordinary differential equation for the actuating force* $f(t)$ *depending on* $d_2(t)$ *can be established as*

$$\mathrm{D}_t f(t) = \frac{1}{3}\,\mathrm{D}_t^3 d_2(t) + \mathrm{D}_t^3\gamma(t)\ . \tag{3.94}$$

By inserting (3.93) *into the remaining differential part of* (3.86), *it becomes apparent, that the concerned function* $d_2(t)$ *is governed by a ordinary differential equation given by*

$$\frac{1}{5}\,\mathrm{D}_t^2 d_2 - d_2 = -\,\mathrm{D}_t^2\gamma(t)\ . \tag{3.95}$$

Eq. (3.95) *can be identified as the internal dynamics of the problem at hand. In this case the internal dynamics are obviously unstable. Subsequently we will try to find the differentially flat output corresponding to the actuation* $f(t)$. *This may lead to a better understanding of the problem at hand and its controllabillity conditions. Following Eq.* (3.90) - *Eq.* (3.92), *a transformation of the state-space variables can be obtained by establishing Kalmans controllability matrix as*

$$\boldsymbol{T}^{-1} = \begin{bmatrix} 0 & 1 & 0 & -5 \\ 0 & 1 & 0 & 15 \\ 1 & 0 & -5 & 0 \\ 1 & 0 & 15 & 0 \end{bmatrix}, \qquad \boldsymbol{\Lambda} = \begin{bmatrix} 0 & 0 & 0 & 0 \\ 1 & 0 & 0 & 0 \\ 0 & 1 & 0 & 15 \\ 0 & 0 & 1 & 0 \end{bmatrix} \qquad \textit{and} \qquad \boldsymbol{Tb} = \begin{bmatrix} 1 \\ 0 \\ 0 \\ 0 \end{bmatrix}$$

The actuating force $f(t)$ *can then be determined as*

$$f(t) = \mathrm{D}_t^4 z_4 - 15\,\mathrm{D}_t^2 z_4.$$

The transformation $\boldsymbol{z} = \boldsymbol{Tx}$ *yields* $z_4(t) = \frac{1}{20}(d_2 - d_1)$, *hence*

$$f = f(\mathrm{D}_t^4 d_1, \mathrm{D}_t^2 d_1, \mathrm{D}_t^4 d_2, \mathrm{D}_t^2 d_2)\ .$$

The function $z_4(t) = \frac{1}{20}(d_2 - d_1)$ *represents a differentially flat output corresponding to the actuation* $f(t)$. *Consequently, in case of not choosing* $z_4(t)$ *as output, (unstable) internal dynamics may occur (cf. Eq.* (3.89)*).*

Example 3.29 (Linear elastic rod, $p = 3$). *By using* $p = 3$ *ansatz functions, the coefficients of the semi-discrete equations of motion are obtained as*

$$\boldsymbol{M} = \begin{bmatrix} 1 & 1/3 & 2/\pi - \pi/6 \\ & 1/5 & -4/\pi^3 + 1/\pi - \pi/20 \\ \mathit{Sym.} & & 1/2 - 8/\pi^2 + \pi^2/30 \end{bmatrix}, \quad \boldsymbol{K} = \begin{bmatrix} 0 & 0 & 0 \\ & 4/3 & \pi/3 - 4/\pi \\ \mathit{Sym.} & & 5\pi^2/6 - 8 \end{bmatrix} \quad \textit{and} \quad \boldsymbol{B}^T = \begin{bmatrix} 1 \\ 1 \\ 0 \end{bmatrix}$$

Again, the state-space variables can be transformed using Kalmans controllability matrix

$$T^{-1} = \begin{bmatrix} 0 & 1 & 0 & a & 0 & d \\ 0 & 1 & 0 & b & 0 & e \\ 0 & 0 & 0 & c & 0 & f \\ 1 & 0 & a & 0 & d & 0 \\ 1 & 0 & b & 0 & e & 0 \\ 0 & 0 & c & 0 & f & 0 \end{bmatrix}, \quad \Lambda = TAT^{-1} = \begin{bmatrix} 0 & 0 & 0 & 0 & 0 & 0 \\ 1 & 0 & 0 & 0 & 0 & 0 \\ 0 & 1 & 0 & 0 & 0 & g \\ 0 & 0 & 1 & 0 & 0 & 0 \\ 0 & 0 & 0 & 1 & 0 & 0 \\ 0 & 0 & 0 & 0 & 1 & 0 \end{bmatrix} \quad \text{and} \quad \boldsymbol{Tb} = \begin{bmatrix} 1 \\ 0 \\ 0 \\ 0 \\ 0 \\ 0 \end{bmatrix},$$

where a, b, c, d, e, f and h are constant values. A dependency of the actuating force $f(t)$ on $z_6 = z_6(d_1, d_2, d_3)$ and its time derivatives can be observed, i.e.

$$f = \mathrm{D}_t^7 z_6 + \mathrm{D}_t^3 z_6 - \mathrm{D}_t^2 z_6$$

Transforming the coordinates back into the original state-space variables indicates that a linear combination of d_1, d_2 and d_3 represents a differentially flat output.

$$f = f(\mathrm{D}_t^7 d_1, \mathrm{D}_t^7 d_2, \mathrm{D}_t^7 d_3, \mathrm{D}_t^3 d_1, \mathrm{D}_t^3 d_2, \mathrm{D}_t^3 d_3, \mathrm{D}_t^2 d_1, \mathrm{D}_t^2 d_2, \mathrm{D}_t^2 d_3)$$

A full paramterization of the actuating force $f(t)$ by only using $d_1(t)$ and its time derivatives is likewise not possible.

4. Simultaneous space-time integration

Due to the highly restrictive applicability of solving the inverse dynamics problem sequentially in space and time, this chapter aims to analyze the initial boundary value problem in more detail. In particular, by exposing the underlying hyperbolic structure of the governing partial differential equations it is anticipated to gain more insights into the problem at hand. Enlighting resulting mechanisms of the flow of informations within continuous structures will pave the way to a numerically stable integration of the inverse dynamics problem at hand. In Sec. 4.1 therfore, the initial boundary value problem that has been postulated in Sec. 1.1, is analyzed in more detail. The new insights will motivate to develop in Sec. 4.1 and Sec. 4.2 novel numerical methods that are based on a simultaneous space-time discretization of the governing equations. In Sec. 4.3 numerical examples, including mechanical systems, that already have been introduced in Sec. 2.1, Sec. 2.2 and Sec. 2.3, are presented underpinning the relevance of the proposed methods.

4.1. Wave propagation

Spatially 1-dimensional systems. Considering 1-dimensional systems, the wave equation for the control problem (1.1), that has been introduced in Sec. 1.1, can be transformed into a system of first-order partial differential equations

$$\begin{aligned} \boldsymbol{A}\partial_t \boldsymbol{v} - \partial_s(\boldsymbol{B}\boldsymbol{p}) &= \boldsymbol{C} \,, \\ \boldsymbol{B}\partial_t \boldsymbol{p} - \boldsymbol{B}\partial_s \boldsymbol{v} &= \boldsymbol{0} \end{aligned} \tag{4.1}$$

by defining the functions $\boldsymbol{v}(s,t) \equiv \partial_t \boldsymbol{x}(s,t)$ and $\boldsymbol{p}(s,t) \equiv \partial_s \boldsymbol{x}(s,t)$. Making use of the product rule, i.e. $\boldsymbol{B}\partial_t \boldsymbol{p} = \partial_t(\boldsymbol{B}\boldsymbol{p}) - \partial_t \boldsymbol{B}\boldsymbol{p}$, Eq. $(4.1)_2$ is obtained as

$$\partial_t(\boldsymbol{B}\boldsymbol{p}) - \boldsymbol{B}\partial_s \boldsymbol{v} = \partial_t \boldsymbol{B}\boldsymbol{p} \,. \tag{4.2}$$

By taking advantage of the *equality of mixed partials*[1], hence $\partial_t \boldsymbol{p} = \partial_s \boldsymbol{v}$ as well as using the chain rule for

$$\partial_t \boldsymbol{B}(\boldsymbol{p}(s,t)) = (\partial_{\boldsymbol{p}} \otimes \boldsymbol{B}) \cdot \partial_t \boldsymbol{p} = \mathrm{grad}_{\boldsymbol{p}}(\boldsymbol{B}) \cdot \partial_t \boldsymbol{p} \,,$$

the following relation for $(4.1)_2$ can be established

$$\partial_t(\boldsymbol{B}\boldsymbol{p}) - \boldsymbol{B}\partial_s \boldsymbol{v} = (\partial_{\boldsymbol{p}} \otimes \boldsymbol{B})\boldsymbol{p}\partial_s \boldsymbol{v} \,. \tag{4.3}$$

Then Eq. $(4.1)_2$ and Eq. (4.3) are forming a system of first-order partial differential equations given by

$$\begin{bmatrix} \boldsymbol{A} & \boldsymbol{0} \\ \boldsymbol{0} & \boldsymbol{I} \end{bmatrix} \begin{bmatrix} \boldsymbol{v} \\ \boldsymbol{B}\boldsymbol{p} \end{bmatrix}_{,t} - \begin{bmatrix} \boldsymbol{0} & \boldsymbol{I} \\ \boldsymbol{H} & \boldsymbol{0} \end{bmatrix} \begin{bmatrix} \boldsymbol{v} \\ \boldsymbol{B}\boldsymbol{p} \end{bmatrix}_{,s} = \begin{bmatrix} \boldsymbol{C} \\ \boldsymbol{0} \end{bmatrix} \,. \tag{4.4}$$

In Eq. (4.4), the function $\boldsymbol{H} : \Omega \mapsto \mathbb{R}^{d,d}$ has been defined as

$$\boldsymbol{H}(\boldsymbol{p}) \equiv \boldsymbol{B} + (\partial_{\boldsymbol{p}} \otimes \boldsymbol{B})\boldsymbol{p} \,. \tag{4.5}$$

By introducing the column vectors $\boldsymbol{z} = [\boldsymbol{v} \ \boldsymbol{B}\boldsymbol{p}]^T : \Omega \mapsto \mathbb{R}^{2d}$ and $\boldsymbol{F} : \Omega \mapsto \mathbb{R}^{2d}$ as well as the square matrices $\boldsymbol{D} : \Omega \mapsto \mathbb{R}^{2d,2d}$ and $\boldsymbol{E} : \Omega \mapsto \mathbb{R}^{2d,2d}$, Eq. (4.4) might be written more compactly as

$$\boldsymbol{D}\partial_t \boldsymbol{z} + \boldsymbol{E}\partial_s \boldsymbol{z} = \boldsymbol{F} \,. \tag{4.6}$$

Assume there exists a line $s = k(t)$ along that a solution $\boldsymbol{z} = \boldsymbol{z}(k(t), t) = \boldsymbol{z}_0(t)$ is given. Then this line is called a characteristic line if the partial derivatives of the solution cannot be uniquely determined through given informations along this line, i.e. Eq. (4.6) along with

$$\partial_t \boldsymbol{z} + \partial_s \boldsymbol{z} \frac{d}{dt} k(t) = \frac{d}{dt} \boldsymbol{z}_0(t) \,, \tag{4.7}$$

hence

$$\left[\boldsymbol{E} - \boldsymbol{D}\frac{d}{dt}k(t)\right] \partial_s \boldsymbol{z} = \boldsymbol{F} - \boldsymbol{D}\frac{d}{dt}\boldsymbol{z}_0(t) \tag{4.8}$$

cannot be solved uniquely for the partial derivatives $\partial_s \boldsymbol{z}$ and $\partial_t \boldsymbol{z}$. If the determinant of the coefficient matrix

$$\boldsymbol{Q} = \boldsymbol{E} - \boldsymbol{D}\frac{d}{dt}k(t) \tag{4.9}$$

as well as the determinants of the matrices, defined by exchanging columns by the right-hand side of Eq. (4.8), vanish then according to *Cramer's rule*, Eq. (4.6) can be transformed into a system of ordinary differential equations along characteristic lines, viz.

$$\det(\boldsymbol{Q}) = 0 \qquad \text{and} \qquad \det_j(\boldsymbol{Q}) = 0 \,, \tag{4.10}$$

[1] also frequently referred to as *Schwarz' theorem* (see e.g. [8, Chap. 24.1 pp. 879-881]).

where $\det_j(Q)$ denotes the determinant of Q where the the j-th column of Q is replaced by the right hand-side of Eq. (4.8), e.g. the components of the j-th column of Q are given by $Q_i = \left(F - D\frac{d}{dt}z_0(t)\right)_i$.

Eq. $(4.10)_1$ and Eq. $(4.10)_2$ are referred to as directionality and compatibility condition, respectively. The procedure outlined above, is also referred to as *Method of Characteristics* ([1, 31, 33, 57, 84]). The resulting system of ordinary differential equations can then be solved either (if possible) analytically or graphically-numerically (cf. Rem. 4.8).

For a more comprehensive geometrical interpretation of the underlying structure, see Rem. 4.12 and Rem. 4.13 as well as the accompanying examples regarding both (quasi-) linear (Ex. 4.14) and general non-linear (Ex. 4.15 and Ex. 4.16) partial differential equations.

Spatially α-dimensional systems. In principle the same holds true for spatially higher dimensional systems, i.e. spatial variables $s \in S \subset \mathbb{R}^\alpha$ for $\alpha > 1$. Introducing $P \equiv B \cdot \partial_s x$, the second term of the partial differential equation (2.11), introduced in Chap. 2 is the divergence of $P(s,t)$ as given by

$$\operatorname{div}_s P = \operatorname{div}_s(B \cdot \partial_s x) = \partial_{s_k}(B_{ij}\partial_{s_k}x_j) = \partial_{s_1}(B_{ij}\partial_{s_1}x_j) + \cdots + \partial_{s_n}(B_{ij}\partial_{s_n}x_j)\ . \tag{4.11}$$

Then Eq. (1.1) takes the form

$$A\partial_t^2 x - \left(\partial_{s_1}(B\partial_{s_1}x) + \cdots + \partial_{s_n}(B\partial_{s_n}x)\right) = C\ . \tag{4.12}$$

By introducing $v(s,t) \equiv \partial_t x$ and $p(s,t) \equiv \partial_s x$ and making use of the *equality of mixed partials*, i.e. $\partial_t p = \partial_s v$ the partial differential equation (4.12) can be transformed into a system of first-order partial differential equations given by

$$D\partial_t z - \left(\partial_{s_1}(E_1 z) + \cdots + \partial_{s_n}(E_n z)\right) = F\ . \tag{4.13}$$

In analogy to the spatially *one*-dimensional case, characteristics can be found, by searching manifolds in the space-time domain on which initial data leads to an indifferent solution $z = \begin{bmatrix} v & p \end{bmatrix}^T$ of the initial boundary value problem. Consequently the inner derivatives

$$\begin{aligned}
\partial_{s_1} z &= \mathrm{D}_{s_1} z - (\partial_t z)(\mathrm{D}_{s_1} t)\ ,\\
&\vdots\\
\partial_{s_n} z &= \mathrm{D}_{s_n} z - (\partial_t z)(\mathrm{D}_{s_n} t)
\end{aligned}$$

give rise to the following system of equations

$$\left[D - \left(E_1\frac{\mathrm{d}s_1}{\mathrm{d}t} + \cdots + E_n\frac{\mathrm{d}s_n}{\mathrm{d}t}\right)\right]\partial_t z = F - \left[E_1\frac{\mathrm{d}z}{\mathrm{d}s_1} + \cdots + E_n\frac{\mathrm{d}z}{\mathrm{d}s_n}\right].$$

By evaluating the determinants of

$$Q = D - (E_1\frac{\mathrm{d}s_1}{\mathrm{d}t} + \cdots + E_n\frac{\mathrm{d}s_n}{\mathrm{d}t})\ ,$$

the system of partial differential equations (4.13) can be transformed into a system of ordinary differential equations (4.10) along characteristic manifolds, i.e. waves can be observed, travelling in *three-* and *four*-dimensional space-time domain along characteristic *surfaces* ($\alpha = 2$) and *volumina* ($\alpha = 3$), respectively. In Ex. 4.1, Ex. 4.4 and Ex. 4.7 (linear case) and Ex. 4.2 and Ex. 4.5 (quasi-linear case) the wave propagation of mechanical systems, that have been introduced in Chap.2, will be analyzed.

Example 4.1 (Wave propagation for the linear elastic bar formulation). *The linear-elastic bar in one dimension ($d = 1$) has been introduced in Ex. 2.5. Following Sec. 4.1, the partial differential equation* (2.14) *can be replaced by the system of first-order partial differential equations* (4.6) *based on the state vector*

$$\boldsymbol{z} = \begin{bmatrix} v \\ n \end{bmatrix}$$

along with

$$\boldsymbol{D} = \begin{bmatrix} \rho A & 0 \\ 0 & 1 \end{bmatrix} \qquad \boldsymbol{E} = -\begin{bmatrix} 0 & 1 \\ EA & 0 \end{bmatrix} \qquad \boldsymbol{F} = \begin{bmatrix} 0 \\ 0 \end{bmatrix} . \tag{4.14}$$

The characteristics are then obtained as

$$\frac{ds}{dt} = k'(t) = \pm c \, . \tag{4.15}$$

Accordingly, the slope of the characteristic lines in the space-time domain Ω is determined by the constant velocity of wave propagation, already introduced in (2.16). *In addition to condition* $(4.10)_1$, *problem* $(4.10)_2$ *gives rise to two further conditions*

$$\det \begin{bmatrix} \rho A \frac{d}{dt}\hat{v} & 1 \\ \frac{d}{dt}\hat{n} & k'(t) \end{bmatrix} = 0 \quad \textit{and} \quad \det \begin{bmatrix} k'(t)\rho A & \rho A \frac{d}{dt}\hat{v} \\ EA & \frac{d}{dt}\hat{n} \end{bmatrix} = 0 \, . \tag{4.16}$$

Here, $\hat{v}(t) = v(k(t), t)$ and $\hat{n}(t) = n(k(t), t)$ stand for, respectively, the velocity and the normal force along the characteristic lines $s = k(t)$ whose slope $k'(t)$ is given by (4.15). *The two conditions in* (4.16) *eventually yield one independent equation of the form*

$$k'(t)\rho A \frac{d\hat{v}}{dt} - \frac{d\hat{n}}{dt} = 0 \, . \tag{4.17}$$

The system of ordinary differential equations (4.17) *can be solved along the characteristic lines specified by* (4.15). *Alltogether, application of the method of characteristics converts the scalar second-order partial differential equation* (2.14) *into a system of four ordinary differential equations given by* (4.15) *and* (4.17). *These ordinary differential equations are supplemented with boundary conditions*

$$n(0, t) = f_0(t), \quad v(L, t) = \gamma'(t), \quad n(L, t) = 0 \tag{4.18}$$

and initial conditions $v(s, 0) = v_0$ and $n(s, 0) = 0$. For the linear problem at hand, Eq. (4.15) *and Eq.* (4.17) *are decoupled. Accordingly, Eq.* (4.15) *can be used in a first step to set up* *the characteristic net consisting of straight characteristic lines specified by the constant velocity of wave propagation c. A sample characteristic net thus obtained is shown in Ex. 4.23, Fig. 4.22. To each node* $(s_Q, t_Q) \in \Omega$ *of the characteristic net there is associated a nodal state vector* $z_Q = (v_Q, n_Q)$*. The unknown nodal states can be determined by integrating Eq.* (4.17) *along the characteristic lines. This procedure is illustrated with Fig. 4.1 in which three nodes* Q *and* P_j, $j \in \{1, 2\}$ *are considered. The three nodes correspond to vertices on the characteristic net and thus comply with Eq.* (4.15) *in the sense that*

$$\frac{s_Q - s_{P_j}}{t_Q - t_{P_j}} + (-1)^j c = 0$$

is satisfied for $j \in \{1, 2\}$*. The numerical integration of* (4.17) *relies on the finite difference scheme*

$$\frac{n_Q - n_{P_j}}{t_Q - t_{P_j}} + (-1)^j c\rho A \frac{v_Q - v_{P_j}}{t_Q - t_{P_j}} = 0$$

for $j \in \{1, 2\}$*. Repeatedly applying this procedure to the characteristic net in Fig. 4.22 yields a (fully coupled) system of algebraic equations for the determination of the unknown states* (v_A, n_A)*. Note that the states at the boundaries at* $s = L$ *(cf. boundary conditions* $(4.18)_{2,3}$*) and* $t = 0$ *(initial conditions) are given. Eventually, boundary condition* $(4.18)_1$ *determines the actuating forces* $f(t_B) = n_B$ *for the nodes B lying on the boundary at* $s = 0$*.*

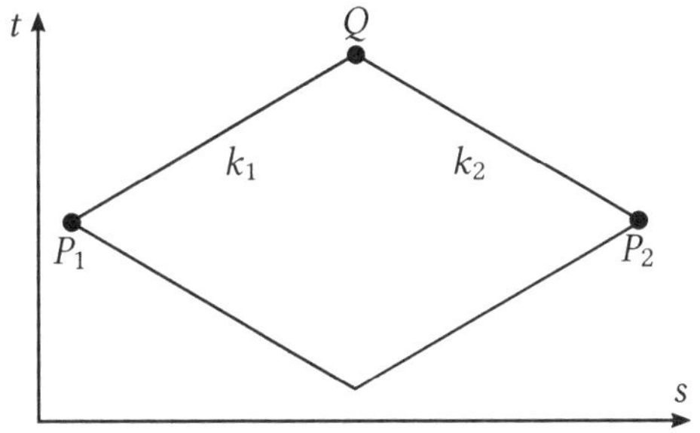

Figure 4.1.: Three nodes Q, P_1 and P_2, located on the characteristic net with associated states (v_Q, n_Q), (v_{P_1}, n_{P_1}) and (v_{P_2}, n_{P_2}). The characteristic lines k_1 and k_2 are associated with the slopes $ds/dt = c$ and $ds/dt = -c$, respectively.

Example 4.2 (Wave propagation for the geometrically exact string formulation). *In line with the procedure outlined in Sec. 4.1, a curve* $s = k(t)$ *along that a solution* $z_0(t) = z(k(t), t)$ *is given, is called a characteristic curve if the partial derivatives* $\partial_s z$ *and* $\partial_s z$ *cannot be determined uniquely by the exclusive knowledge of* z *on curve* $k(t)$*, i.e. the condition*

$$(\boldsymbol{E} - k'(t)\boldsymbol{D})\, \partial_s \boldsymbol{z} = \boldsymbol{F} - \boldsymbol{D}\frac{d}{dt}\widehat{\boldsymbol{z}}(t) \tag{4.19}$$

for

$$\boldsymbol{D} = \begin{bmatrix} A & 0 \\ 0 & I \end{bmatrix} \qquad \boldsymbol{E} = -\begin{bmatrix} 0 & I \\ H & 0 \end{bmatrix} \qquad \boldsymbol{F} = \begin{bmatrix} C \\ 0 \end{bmatrix} \tag{4.20}$$

gives rise to the condition

$$\det(\boldsymbol{E} - k'(t)\boldsymbol{D}) = 0 \,. \tag{4.21}$$

Consequently, $k'(t)$ coincide with the eigenvalues of $\boldsymbol{E}$ relative to $\boldsymbol{D}$ (i.e. the eigenvalues of $\boldsymbol{D}^{-1}\boldsymbol{E}$). Correspondingly, there are 2d characteristics associated with the eigenvalues λ_i for $i \in \{1, \dots, 2d\}$. In particular, for the hyperelastic material model introduced in Ex. 2.1, function $\boldsymbol{H}$ in Eq. $(4.20)_2$ *assumes the form*

$$\boldsymbol{H} = \frac{EA}{2}\left(1 - \frac{1}{\nu^2}\right)\boldsymbol{I} + \frac{EA}{\nu^4}\boldsymbol{p} \otimes \boldsymbol{p} \tag{4.22}$$

so that condition (4.21) *yields the pairwise solution*

$$k'(t) = \pm c\sqrt{\frac{1}{2}\left(1 + \frac{1}{\nu^2}\right)} \tag{4.23}$$

for $d = 1$,

$$k_1'(t) = \pm c\sqrt{\frac{1}{2}\left(1 + \frac{1}{\nu^2}\right)} \quad \text{and} \quad k_2'(t) = \pm c\sqrt{\frac{1}{2}\left(1 - \frac{1}{\nu^2}\right)} \tag{4.24}$$

for $d = 2$ and

$$k_1'(t) = \pm c\sqrt{\frac{1}{2}\left(1 + \frac{1}{\nu^2}\right)} \quad \text{and} \quad k_{2,3}'(t) = \pm c\sqrt{\frac{1}{2}\left(1 - \frac{1}{\nu^2}\right)} \tag{4.25}$$

for $d = 3$. In Eq. (4.23) *- Eq.* (4.25) *c denotes the constant velocity of wave propagation introduced in* (2.16).

Apart from the one-dimensional case ($d = 1$), the eigenvalues (4.24) *and* (4.25) *are not all real-valued anymore for $\nu < 1$. This conforms with the fact that strings can sustain tensile forces but not compressive ones. The problem is then no more hyperbolic.*

Example 4.3 (Plane wave propagation for the geometrically exact string formulation). *For the nonlinear string in two dimensions ($d = 2$), the second-order partial differential equation* $(2.11)_1$ *is replaced by the system of first-order partial differential equations* (4.6) *based on the matrices in* (4.20) *and* (4.22). *The state vector $\boldsymbol{z} : \Omega \mapsto \mathbb{R}^4$ is comprised of the vectors $\boldsymbol{v} : \Omega \mapsto \mathbb{R}^2$ and $\boldsymbol{n} : \Omega \mapsto \mathbb{R}^2$. Then, two pairs of characteristics can be distinguished which are characterized by the slopes in* (4.24). *Correspondingly, two ($i = 1, 2$) pairs of characteristics are defined by*

$$\left(\frac{ds}{dt}\right)_{ij} = (-1)^{j+1} c_i \tag{4.26}$$

where, with regard to (4.24),

$$c_1 = c\sqrt{\frac{1}{2}\left(1 + \frac{1}{\nu^2}\right)} \quad \text{and} \quad c_2 = c\sqrt{\frac{1}{2}\left(1 - \frac{1}{\nu^2}\right)} \,. \tag{4.27}$$

Each pair of characteristics has $j \in \{1,2\}$ members, where $j = 1$ refers to the forward propagating direction and $j = 2$ to the backward propagating direction. In addition to the directionality condition (4.26), the compatibility condition indicates a system of ordinary differential equations along the characteristics $i, j \in \{1,2\}$ given by

$$\boldsymbol{U}_i \cdot \left(\frac{d\widehat{\boldsymbol{n}}}{dt}\right)_i + (-1)^j c_i \boldsymbol{V}_i \cdot \left(\frac{d\widehat{\boldsymbol{v}}}{dt}\right)_i + (-1)^j c_i W = 0 \tag{4.28}$$

In Eq. (4.28), the functions $\widehat{\boldsymbol{v}} = \boldsymbol{v}(k(t), t)$ and $\widehat{\boldsymbol{n}} = \boldsymbol{n}(k(t), t)$ denote, respectively, the velocity and the contact force along the respective characteristic curve. Moreover,

$$\boldsymbol{U}_i = \boldsymbol{V}_i = \boldsymbol{P}_i\, \boldsymbol{p} \qquad \textit{and} \qquad W = p_1 p_2$$

where

$$\boldsymbol{P}_i = \delta_{i1} \begin{bmatrix} p_1 & 0 \\ 0 & p_1 \end{bmatrix} + \delta_{i2} \begin{bmatrix} 0 & -p_2 \\ p_2 & 0 \end{bmatrix} .$$

Equations (4.26) and (4.28) constitute a system of coupled non-linear ordinary differential equations which can be discretized as follows:

$$\begin{aligned} \frac{s_Q - s_{P_{ij}}}{t_Q - t_{P_{ij}}} + (-1)^j c_i\big|_{P_{ij}} &= 0 , \\ \boldsymbol{U}_i\big|_{P_{ij}} \frac{\boldsymbol{n}_Q - \boldsymbol{n}_{P_{ij}}}{t_Q - t_{P_{ij}}} + (-1)^j (c_i \boldsymbol{V}_i)\big|_{P_{ij}} \frac{\boldsymbol{v}_Q - \boldsymbol{v}_{P_{ij}}}{t_Q - t_{P_{ij}}} + (-1)^j (c_i W)\big|_{P_{ij}} &= 0 . \end{aligned} \tag{4.29}$$

In analogy to Ex. 4.1, vertices of the characteristic net are denoted by (s_Q, t_Q), while the corresponding nodal states are denoted by $(\boldsymbol{v}_Q, \boldsymbol{n}_Q)$. In contrast to Ex. 4.1, the characteristic net can not be determined beforehand, because c_i in $(4.29)_1$ depends on the state, see (4.27). Consequently, the location of the nodal net points, (s_Q, t_Q) and $(s_{P_{ij}}, t_{P_{ij}})$, in general need to be determined through the solution procedure, as well as the nodal states.

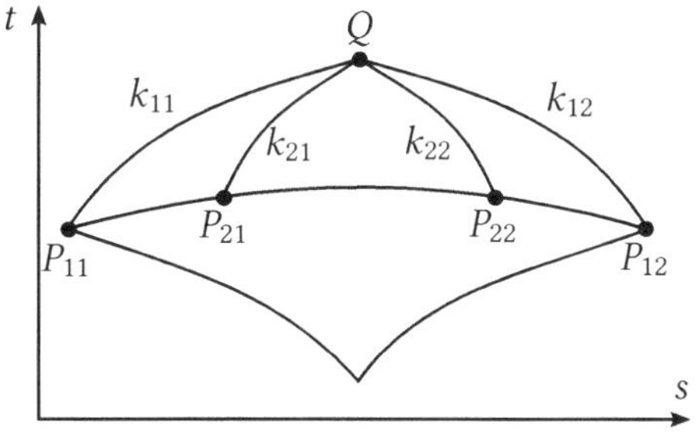

Figure 4.2.: Integration along the characteristics k_{ij} for $i, j \in \{1,2\}$ according to (4.29). The characteristics k_{ij} are associated with the slopes $(ds/dt)_{ij}$ in (4.26).

Scheme (4.29) is further illustrated with Fig. 4.2, in which the points Q and P_{ij} located on the respective characteristics are displayed. This procedure can be repeatedly applied within *the space-time domain of interest. After assembly, a coupled system of non-linear algebraic equations is obtained which can be solved iteratively by means of Newton's method. Prior to the solution, the boundary and initial conditions emanating from* $(2.11)_{2,3}$ *need to be taken into account. Rem. 4.8 contains an outline of the implementation based on scheme* (4.29).

Example 4.4 (Wave propagation for the linear Timoshenko-Ehrenfest beam formulation). *Consulting Rem. 2.14, the motion of a beam, formulated in terms of the Timoshenko-Ehrenfest formulation is governed by a system of second-order partial differential equations given by*

$$\rho A \partial_t^2 u_1 - \partial_s(H) - \bar{n} = 0 \, ,$$
$$\rho I \partial_t^2 \Theta - \partial_s(M) + H - \bar{m} = 0 \, .$$

Regarding bending and shear deformations, constitutive relations are assumed as

$$M = EI\partial_s\Theta = EI\kappa \quad ; \quad H = GA(\partial_s u_1 + \Theta) = GA\gamma \, .$$

The initial boundary value problem at hand aligns with the framework proposed in Sec. 1.1 for $\boldsymbol{x} = \begin{bmatrix} u_1 & \Theta \end{bmatrix}^T$ *along with*

$$\boldsymbol{A} = \begin{bmatrix} \rho A & 0 \\ 0 & \rho I \end{bmatrix}, \quad \boldsymbol{B} = \begin{bmatrix} GA & 0 \\ 0 & EI \end{bmatrix} \quad \text{and} \quad \boldsymbol{C} = -\begin{bmatrix} \bar{n} + GA\partial_s\Theta \\ H - \bar{m} \end{bmatrix} \, . \tag{4.30}$$

Evaluating the directionality condition $(4.10)_1$ *gives rise to the characteristic directions*

$$c_1 = \pm \left(\frac{E}{\rho}\right)^{\frac{1}{2}} \quad \text{and} \quad c_2 = \pm \left(\frac{G}{\rho}\right)^{\frac{1}{2}} \, .$$

Herein, c_1 *and* c_2 *can be identified as the velocities of bending and shear waves, respectively. In Fig 4.3 the wave propagation for the Timoshenko-Ehrenfest beam formulation is illustrated. Suppose, that* $\overline{P_{11}P_{12}}$ *is an initial curve, along that the solution can be considered as given. Then the solution propagates along characteristic curves* k_{ij} *for* $i, j \in \{1, 2\}$ *such that the solution in point* Q *depends on points* P_{ij} *for* $i, j \in \{1, 2\}$.

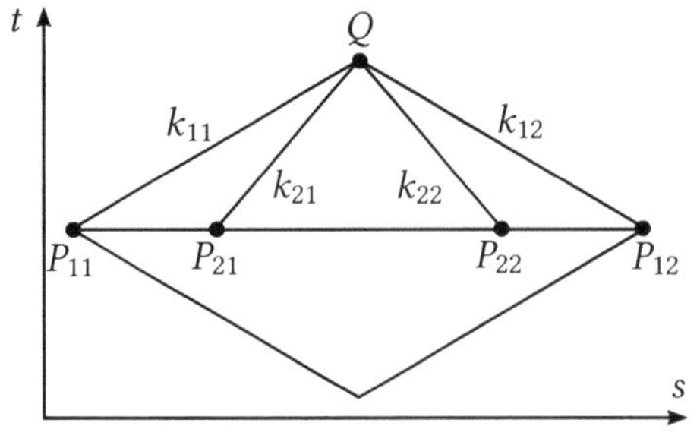

Figure 4.3.: Wave propagation for the linear Timoshenko-Ehrenfest beam formulation.

Example 4.5 (Plane wave propagation for the geometrically exact beam formulation - componential representation). *Regarding Rem. 2.12 and assuming hyperelastic material (cf. Ex. 2.10), the function* $\boldsymbol{H} : \Omega \mapsto \mathbb{R}^{d,d}$ *can be identified as*

$$H_{ik} = B_{ik} + \partial_{p_k} B_{ij} p_j = \begin{bmatrix} \frac{EA}{2}(1+\gamma_3^{-2})\cos\Theta & \frac{EA}{2}(1+\gamma_3^{-2})\sin\Theta & 0 \\ -GA\sin\Theta & GA\cos\Theta & 0 \\ 0 & 0 & EI \end{bmatrix} . \tag{4.31}$$

where, use of

$$\partial_{p_k} B_{ij} p_j = \begin{cases} EA\gamma_3^{-2}\cos\Theta & \text{for} \quad i = k = 1 \\ EA\gamma_3^{-2}\sin\Theta & \text{for} \quad i = 1 \wedge k = 2 \\ 0 & \text{otherwise} \end{cases} \tag{4.32}$$

has been made. Note that, according to Ex. 2.8, the strain variable γ_3 *is given by the relation* $\gamma_3 = \partial_s r_1 \cos\Theta + \partial_s r_2 \sin\Theta$. *By evaluating the directionality condition* $(4.10)_1$, *the following characteristic slopes can be identified:*

$$c_1 = \pm\left(\frac{EI}{\rho I}\right)^{\frac{1}{2}}, \quad c_2 = \pm\left(\frac{GA}{\rho A}\right)^{\frac{1}{2}} \quad \text{and} \quad c_3 = \pm\left(\frac{1}{2}\frac{EA}{\rho A}\left(1 + \frac{1}{\gamma_3^2}\right)\right)^{\frac{1}{2}} \tag{4.33}$$

Herein, c_1, c_2 *and* c_3 *can be interpreted as the velocity of wave propagation corresponding to bending, shear and compression, respectively.*

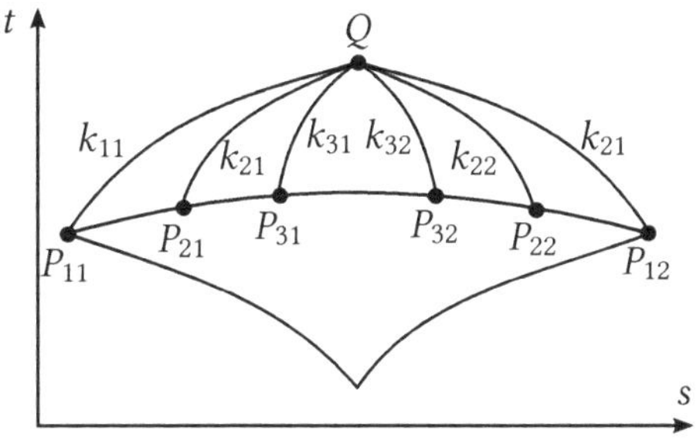

Figure 4.4.: Wave propagation for the Cosserat beam formulation for nonlinear constitutive relations.

In addition to that, the compatibility condition indicates further ordinary differential equations, governing the solution along characteristic lines. Together, a system of ordinary differential equations is obtained that is equivalent to the underlying partial differential equation. This enables an integration of the problem at hand globally in space-time domain Ω *(cf. [31]). Observing that* $\partial_s \boldsymbol{r} = \|\partial_s \boldsymbol{r}\| \begin{bmatrix}\cos\alpha & \sin\alpha & 0\end{bmatrix}^T$, *the strain variable* γ_3 *might be reshaped as*

$$\begin{aligned} \gamma_3 = \partial_s r_1 \cos\Theta + \partial_s r_2 \sin\Theta &= (\cos\alpha\cos\Theta + \sin\alpha\sin\Theta)\,\|\partial_s r\| \\ &= \|\partial_s \boldsymbol{r}\|\cos(\alpha - \Theta) = \|\partial_s \boldsymbol{r}\|\cos\beta \,. \end{aligned}$$

Therein, use of the identities

$$\cos(\alpha - \Theta) + \cos(\alpha + \Theta) = 2\cos\alpha\cos\Theta\,,$$
$$\cos(\alpha - \Theta) - \cos(\alpha + \Theta) = 2\sin\alpha\sin\Theta$$

has been made. See Fig. 4.5 for a geometrical interpretation. In case of assuming $\beta = 0$ the velocity of longitudinal wave propagation of the geometrically exact beam formulation coincides with the velocity of longitudinal wave propagation of the geometrically exact string formulation (cf. Eq. (4.27) in Ex. 4.2).

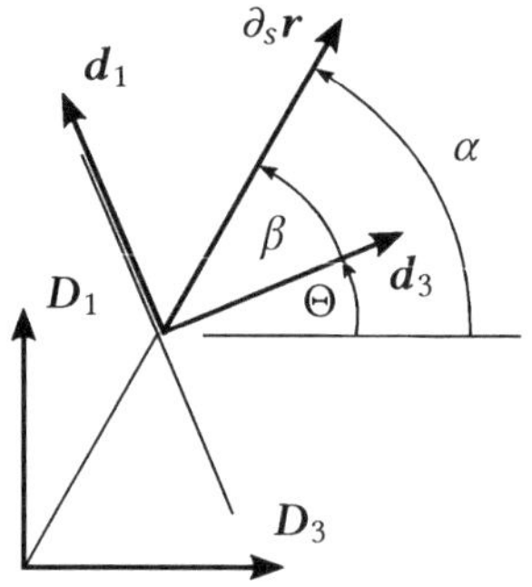

Figure 4.5.: Planar illustration of the orientation of a cross-section of a geometrical exact beam.

See Fig. 4.6 and Fig. 4.7 for an graphical evaluation of the longitudinal velocity c_3 for the geometrically exact beam formulation depending on the shear angle β for different $\|\partial_s r\| \in \{0.5, 1.1, 1.4, 2.0, 5.0\}$ and depending on $\|\partial_s r\|$ for $\beta \in \{0, \frac{\pi}{4},, \frac{\pi}{3}, \frac{2}{3}\pi, \frac{3}{4}\pi, \pi\}$, respectively.

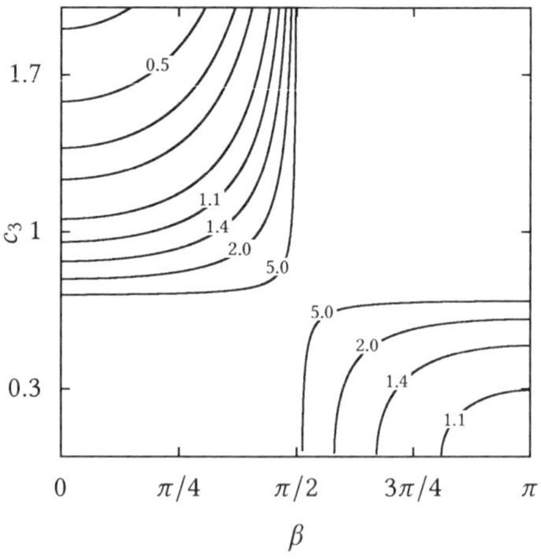

Figure 4.6.: Longitudinal wave propagation c_3 for the geometrically exact beam formulation, depending on β.

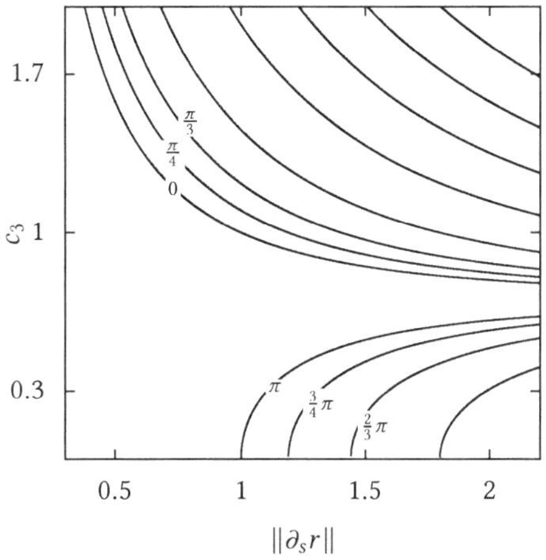

Figure 4.7.: Longitudinal wave propagation c_3 for the geometrically exact beam formulation, depending on $\|\partial_s r\|$.

Example 4.6 (Objectivity). *The velocities of the plane wave propagation for the geometrically exact beam formulation, which have been evaluated in Ex. 4.5, of course need to be independent to any coordinate transformation. Rotating the frame of reference, e.g. by* $\Lambda \in SO(3)$, *must not affect the characteristic directions. Hence, changing from componential to vectorial representation must lead to the same velocities of wave propagation. Thus, regarding the formulation in vectorial representation and thereby taking the coefficients from Ex. 2.11, the matrix* $\boldsymbol{H}$ *can be determined as*

$$H_{ik} = B_{ik} + \partial_{p_k} B_{ij} p_j = \frac{EA}{2}(1 + \gamma_3^{-2}) \begin{bmatrix} \cos^2\Theta & \cos\Theta\sin\Theta & 0 \\ \cos\Theta\sin\Theta & \sin^2\Theta & 0 \\ 0 & 0 & 0 \end{bmatrix} \tag{4.34}$$

$$+ GA \begin{bmatrix} \sin^2\Theta & -\cos\Theta\sin 2\Theta & 0 \\ -\cos\Theta\sin 2\Theta & \cos^2\Theta & 0 \\ 0 & 0 & 0 \end{bmatrix} + EI \begin{bmatrix} 0 & 0 & 0 \\ 0 & 0 & 0 \\ 0 & 0 & 1 \end{bmatrix} .$$

In Eq. (4.34), *the components of* $\operatorname{grad}_p(\boldsymbol{B})\boldsymbol{p}$ *are obtained as*

$$\partial_{p_k} B_{ij} p_j = \begin{cases} EA\gamma_3^{-2}\cos^2\Theta & \textit{for} \quad i = k = 1 \\ EA\gamma_3^{-2}\sin^2\Theta & \textit{for} \quad i = 2 \textit{ and } k = 2 \\ EA\gamma_3^{-2}\cos\Theta\sin\Theta & \textit{for} \quad i = 1 \textit{ and } k = 2 \vee i = 2 \textit{ and } k = 1 \\ 0 & \textit{otherwise.} \end{cases} \tag{4.35}$$

where use of $\gamma_3 = \partial_s r_1 \cos\Theta + \partial_s r_2 \sin\Theta$ *has been made (cf. Ex. 2.8). By evaluating the directionality condition, the velocities of wave propagation are obtained as*

$$c_1 = \pm\left(\frac{EI}{\rho I}\right)^{\frac{1}{2}}, \quad c_2 = \pm\left(\frac{GA}{\rho A}\right)^{\frac{1}{2}} \quad \textit{and} \quad c_3 = \pm\left(\frac{1}{2}\frac{EA}{\rho A}\left(1 + \frac{1}{\gamma^2}\right)\right)^{\frac{1}{2}} . \tag{4.36}$$

Indeed a comparison of (4.36) *with the velocities of wave propagation for the componential representation evaluated in Ex. 4.5 proves the objectivity.*

Example 4.7 (Plane wave propagation for the linear elastic continuum formulation). *Subsequently we confine our attention to linear-elasticity. Assuming infinitesimal deformations, the kinematical considerations that have been elaborated in Sec. 2.3 can be linearized in a direction* $\boldsymbol{u}(\mathbf{s}, t)$ *around the reference configuration. Consequently, the geometrical linearization of the Green-Lagrange strain tensor is obtained as*

$$\mathrm{D}_{\alpha}\boldsymbol{E}(\boldsymbol{F}(\mathbf{s}+\alpha\boldsymbol{u}(\mathbf{s},t)))|_{\alpha=0} = \frac{1}{2}\left(\partial_{\mathbf{s}}\boldsymbol{u}+(\partial_{\mathbf{s}}\boldsymbol{u})^{T}\right) \equiv \boldsymbol{\varepsilon} \tag{4.37}$$

Furthermore, assuming linear isotropy, the Cauchy stress tensor field $\boldsymbol{\sigma}(\boldsymbol{r}, t)$ *can be set constitutively into relation to the linearized strain tensor field* $\boldsymbol{\varepsilon}(\boldsymbol{r}, t)$ *as given by*

$$\boldsymbol{\sigma} = \lambda(\boldsymbol{I} : \boldsymbol{\varepsilon})\boldsymbol{I} + 2\mu\boldsymbol{\varepsilon} \tag{4.38}$$

giving rise to

$$\boldsymbol{\sigma} = \lambda(\boldsymbol{I} : \partial_{\mathbf{s}}\boldsymbol{u})\boldsymbol{I} + \mu(\partial_{\mathbf{s}}\boldsymbol{u} + (\partial_{\mathbf{s}}\boldsymbol{u})^{T}) \ . \tag{4.39}$$

In Eq. (4.38) *the two Lamé parameters* λ *and* μ *have been introduced. Inserting the divergence of* (4.39) *with respect to* $\mathbf{s}$, *i.e.*

$$\operatorname{div}\boldsymbol{\sigma} = \operatorname{div}(\lambda(\boldsymbol{I} : \partial_{\mathbf{s}}\boldsymbol{u})\boldsymbol{I}) + \operatorname{div}(\mu(\partial_{\mathbf{s}}\boldsymbol{u} + (\partial_{\mathbf{s}}\boldsymbol{u})^{T}) \tag{4.40}$$

into the linearized balance of linear momentum

$$\operatorname{div}\boldsymbol{\sigma} + \boldsymbol{f} = \rho\partial_t^2\boldsymbol{u} \ ,$$

a partial differential equation that governs the ininitesimal displacements $\boldsymbol{u}(\mathbf{s}, t)$ *of all material points* $\mathbf{s}$ *at any time* t *is obtained as*

$$(\lambda+\mu)\partial_{\mathbf{s}}(\partial_{\mathbf{s}}\boldsymbol{u}) + \mu\Delta\boldsymbol{u} + \boldsymbol{f} = \rho\partial_t^2\boldsymbol{u} \ . \tag{4.41}$$

These differential equations are frequently referred to as 'Lamés differential equations' or 'Naviers differential equations'. Subsequently, these equations will be analyzed regarding the propagation of waves. Therefore, we restrict ourself to the planar case of linear elasticity, i.e. plane stress and plane strain is assumed, respectively. For more details on wave phenomena of solids we refer to [2] and [61]. For the sake of clarity componential notation is used subsequently.

(i) *Plane stress. Considering the special case of plane stress by assuming* $\sigma_{i3} = \sigma_{3i} = 0$ *for* $i \in \{1, 2, 3\}$ *and* $\varepsilon_{j3} = \varepsilon_{3j} = 0$ *for* $j \in \{1, 2\}$, *the components of the Cauchy stress tensor are obtained as*

$$\sigma_{ij} = \lambda\varepsilon_{kk} + 2\mu\varepsilon_{ij}$$

Consequently, a relation for ε_{33} *depending on* ε_{11} *and* ε_{22} *is given by*

$$\varepsilon_{33} = -\frac{\lambda}{2\mu+\lambda}(\varepsilon_{11}+\varepsilon_{22}) \ .$$

This in turn leads to a relation for the components of the Cauchy stress tensor depending on the linear strain components

$$\sigma_{ij} = \lambda(1 - \frac{\lambda}{2\mu + \lambda})\varepsilon_{kk} + 2\mu\varepsilon_{ij} \ \forall \ i, j, k \in \{1, 2\}$$

Consequently the divergence of the Cauchy stress tensor field yields

$$\partial_j \sigma_{ij} = \lambda \left(1 - \frac{\lambda}{2\mu + \lambda}\right) \begin{bmatrix} \partial_1(\varepsilon_{11} + \varepsilon_{22}) \\ \partial_2(\varepsilon_{11} + \varepsilon_{22}) \end{bmatrix} + 2\mu \begin{bmatrix} \partial_1\varepsilon_{11} + \partial_2\varepsilon_{12} \\ \partial_1\varepsilon_{21} + \partial_2\varepsilon_{22} \end{bmatrix} .$$

Introducing the new functions $\boldsymbol{p} \equiv \begin{bmatrix} \varepsilon_{11} & \varepsilon_{22} & \gamma_{12} \end{bmatrix}^T$ *and* $\boldsymbol{v} \equiv \begin{bmatrix} \partial_t u_1 & \partial_t u_2 \end{bmatrix}^T$ *and taking advantage of the property of mixed partials, i.e.*

$$\partial_t \varepsilon_{11} - \partial_{s_1} v_1 = 0, \quad \partial_t \varepsilon_{22} - \partial_{s_2} v_2 = 0 \quad \textit{and} \quad \partial_t \gamma_{12} - (\partial_{s_1} v_2 + \partial_{s_2} v_1) = 0 \ ,$$

Naviers equations (4.41) *can be established as a system of first-order partial differential equations*

$$\boldsymbol{A}\partial_{s_1}\boldsymbol{z} + \boldsymbol{B}\partial_{s_2}\boldsymbol{z} + \boldsymbol{C}\partial_t\boldsymbol{z} = \boldsymbol{0} \tag{4.42}$$

governing the solution $\boldsymbol{z} = \begin{bmatrix} \boldsymbol{p} & \boldsymbol{v} \end{bmatrix}^T$. *In Eq.* (4.42) *the coefficients*

$$\boldsymbol{A} = \begin{bmatrix} a + 2\mu & a & 0 & 0 & 0 \\ 0 & 0 & \mu & 0 & 0 \\ 0 & 0 & 0 & 0 & -1 \\ 0 & 0 & 0 & -1 & 0 \\ 0 & 0 & 0 & 0 & 0 \end{bmatrix}, \qquad \boldsymbol{B} = \begin{bmatrix} 0 & 0 & \mu & 0 & 0 \\ a & a + 2\mu & 0 & 0 & 0 \\ 0 & 0 & 0 & -1 & 0 \\ 0 & 0 & 0 & 0 & 0 \\ 0 & 0 & 0 & 0 & -1 \end{bmatrix},$$
$$\boldsymbol{C} = \begin{bmatrix} 0 & 0 & 0 & -\rho & 0 \\ a & 0 & 0 & 0 & -\rho \\ 0 & 0 & 1 & 0 & 0 \\ 1 & 0 & 0 & 0 & 0 \\ 0 & 1 & 0 & 0 & -1 \end{bmatrix} \tag{4.43}$$

along with the constant parameter $a = \lambda(1 - \lambda(2\mu + 1)^{-1})$, *have been introduced. Evaluating the directionality condition*

$$\det(\boldsymbol{C} - \mathbf{A}\mathrm{D}_1\boldsymbol{z} - \mathbf{B}\mathrm{D}_2\boldsymbol{z}) = 0$$

yields two distinct velocities of wave propagation

$$c_T^2 = c_1^2 + c_2^2 = \frac{\mu}{\rho} \quad \textit{and} \quad c_L^2 = c_1^2 + c_2^2 = \frac{1}{\rho}\frac{4\mu(\lambda + \mu)}{\lambda + 2\mu} \ . \tag{4.44}$$

Note that c_L *coincides with the longitudinal velocity of waves in an infinite plate (cf. [61, pp.79-83]).*

(ii) *Plane strain case. For the plane strain case, i.e.* $\varepsilon_{i3} = \varepsilon_{3i} = 0$*, Naviers equations* (4.41) *boil down to*

$$\lambda \begin{bmatrix} \partial_1\varepsilon_{11} + \partial_1\varepsilon_{22} \\ \partial_2\varepsilon_{11} + \partial_2\varepsilon_{22} \end{bmatrix} + 2\mu \begin{bmatrix} \partial_1\varepsilon_{11} \\ \partial_2\varepsilon_{22} \end{bmatrix} + \mu \begin{bmatrix} \partial_2\gamma_{12} \\ \partial_1\gamma_{12} \end{bmatrix} - \rho \begin{bmatrix} \partial_t^2 u_1 \\ \partial_t^2 u_2 \end{bmatrix} = \begin{bmatrix} 0 \\ 0 \end{bmatrix} .$$

The problem can be written as a system of first-order partial differential equations taking the form of (4.42) *along with the coefficients introduced in* (4.43) *by assuming* $a = \lambda$*. The slopes of the characteristics can then be observed to be*

$$c_T^2 = c_1^2 + c_2^2 = \frac{\mu}{\rho} \quad \text{and} \quad c_L^2 = c_1^2 + c_2^2 = \frac{\lambda + 2\mu}{\rho} .$$

Herein, c_T *and* c_L *are representing the velocity of transversal and longitudinal wave propagation*[2]*, respectively (cf. [2, pp. 123-124] [43, pp. 318-320] and [61, pp. 10-15]).*

In Fig. 4.8 an illustration of the characteristic directions, along that longitudinal and transversal waves propagate in terms of planar linear elastodynamics of continua, are depicted for the planar formulation of linear elasticity of continua.

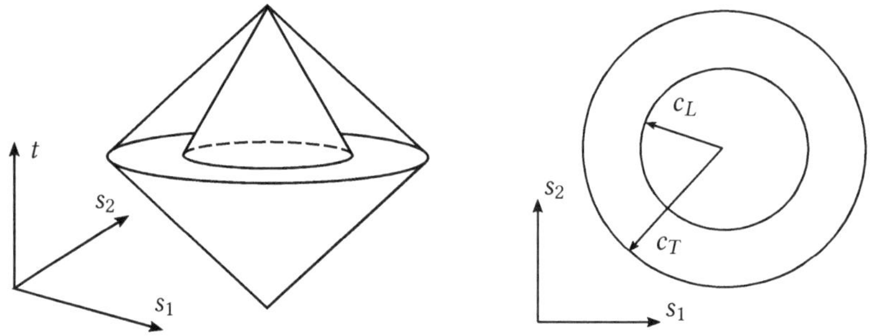

Figure 4.8.: Longitudinal and transversal wave propagation for the linear elastic continuum formulation - the plane stress and plane strain case.

Remark 4.8 (Massaus scheme). *In Sec. 4.1, it has been demonstrated, that the knowledge of characteristic manifolds enables a transformation of hyperbolic partial differential equations into a system of ordinary differential equations. Although ordinary differential equations are much more comfortable to solve than partial differential equations, complex mechanical systems lead to equations that are still challenging, even impossible, to solve analytically. Therefore, a graphically-numerically approach, that could have been implemented, is introduced subsequently.*

Suppose, curve J_0 *is an initial curve, on that the solution* $z : \Omega \mapsto \mathbb{R}^{2d}$ *is considered as given. Then the solution at any point* Q *on curve* J_1 *depends on the solution at points* P_{ij} *and can*

[2] Waves that are traveling with velocity c_L do not involve any rotation and waves that are traveling with velocity c_T do not involve any dilatation. Thus c_L and c_T is frequently referred to as irrotational and equivoluminal velocity of wave propagation, respectively. Alternatively they are referred to as dilatational and distortional waves.

be determined by solving the system of ordinary differential equations (4.10) *along characteristic lines k_{ij} for $i \in \{1, \cdots, d\}$ and $j \in \{1, 2\}$, by applying appropriate finite difference schemes. See Fig. 4.9 for an illustration thereof. Note that curves J_n for $n \in \{0, 1, \cdots\}$ need to be strictly non-characteristic. Once the solution at point Q on curve J_1 has been determined, the solution among major characteristics, i.e. $i = 1$, can be interpolated. In the case of unidirectional wave propagation, interpolation is obviously not required (cf. e.g. Ex. 4.1). The given boundary and initial conditions can be applied directly at the corresponding nodes of the characteristic net. This graphical-numerical approach can be traced back to J. Massau (cf. [72]).*

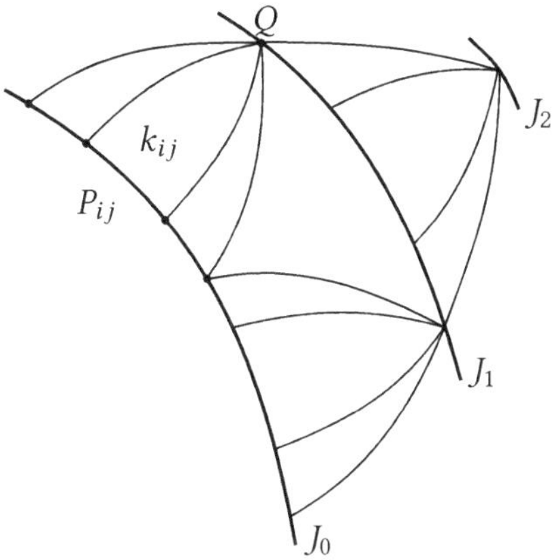

Figure 4.9.: Illustration of the implemented graphical-numerical procedure - Massaus method.

Remark 4.9 (Delay and differential flatness). *Subsequently, some remarks on the delay of perturbations and its connection to the property of differential flatness are made. Therfore, the domain Ω that is bounded by three non-characteristic curves J_0, K_0 and K_2 as well as by the characteristic curve C_4^+ is considered. See Fig. 4.10 for a graphical illustration thereof.*

Delay. Assume the solution on curve $K_2(\xi)$ for $\xi \in [P_2, P_3]$ is perturbed. Then, waves that are travelling along characteristic lines C will affect the solution on curve $K_0(\xi)$ for $\xi \in [P_0, P_5]$. Since the solution on $K_0(\xi)$ for $\xi \in [P_0, P_5]$ in turn affects the solution on curve $K_2(\xi)$ for $\xi \in [P_2, P_4]$, the solution on curve $K_2(\xi)$ for $\xi \in [P_3, P_4]$ is affected by reflecting waves caused by perturbations imposed on curve $K_2(\xi)$ for $\xi \in [P_2, P_3]$. Note that the solution on $K_2(\xi)$ for $\xi \in [P_1, P_2]$ is governed by initial conditions that are considered as given on curve J_0. In the context of the inverse dynamics problem, we refer to curve $K_2(\xi)$ for $\xi \in [P_1, P_2]$ as pre-actuation phase and to curve $K_2(\xi)$ for $\xi \in [P_3, P_4]$ as post-actuation phase.

The span of both pre- and post-actuation phases are determined by the slope of the characteristic curves together with the distance between the boundaries. The shaded domain in Fig. 4.10 elucidates this correlation. (See also Ex. 4.23, Fig. 4.22 for the corresponding linear problem.)

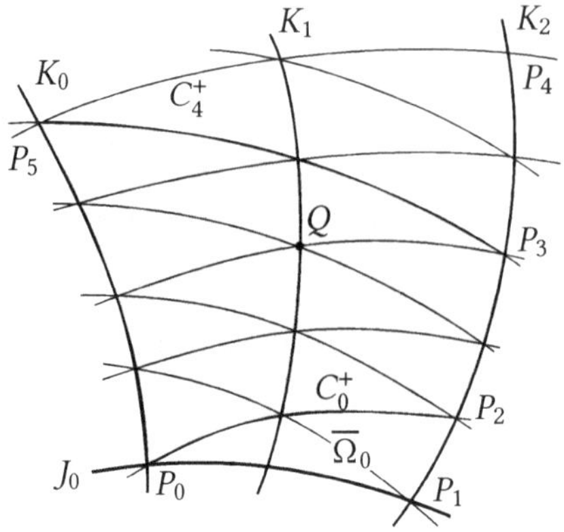

Figure 4.10.: Characteristic net.

Flatness. Concentrating on the space-time tube between the two characteristic curves C_0^+ and C_4^+ the solution $z(K_1)$ at all points on curve K_1 depend on the solution $z(K_2)$ at points on K_2. Once the solution $z(K_1)$ at all points on K_1 has been determined, the solution $z(K_0)$ on K_0 can be established by repeating this procedure. Thus, the solution $z(K_0)$ on K_0 can be totally parameterized by the solution $z(K_2)$ given on K_2.

This is in analogy to the property of differential flatness introduced in Def. 3.23 and accompanying Ex. 3.24, Ex. 3.25 and Ex. 3.26.

Remark 4.10 (Riemann Invariants). *In general, the solution $\xi \in \mathbb{R} \mapsto \boldsymbol{z}(\xi)$ itself is not conserved along characteristic lines. But there might exist quantities, that are preserved along these lines. This remark aims to reveal such invariant functions.*

A function $\boldsymbol{z} \in \mathbb{R}^{2d} \mapsto h(\boldsymbol{z}) \in \mathbb{R}^d$ is a preserved quantity along characteristic lines if

$$\partial_\xi \boldsymbol{h}(\boldsymbol{z}(\xi)) = 0 \tag{4.45}$$

holds. Together with equation (4.6)

$$\partial_\xi \boldsymbol{z} = \partial_t \boldsymbol{z}\, \mathrm{D}_\xi t + \partial_s \boldsymbol{z}\, \mathrm{D}_\xi s$$

and setting therein for simplicity $\mathbf{F} = \mathbf{0}$ the following relation can be established:

$$\left(-(\mathbf{D}^{-1}\mathbf{E})^T \partial_{\boldsymbol{z}} \boldsymbol{h} + c_i \partial_{\boldsymbol{z}} \boldsymbol{h}\right) \partial_s \boldsymbol{z} = \mathbf{0} \; .$$

Thus, the eigenvectors $\partial_{\boldsymbol{z}} \boldsymbol{h}$ associated to the eigenvalues of $\mathbf{D}^{-1}\mathbf{E}$ yield along characteristic lines the preserved quantity $\boldsymbol{h}(\boldsymbol{z})$. The following eigenvalue problem can then be established:

$$(\mathbf{D}^{-1}\mathbf{E})^T \partial_{\boldsymbol{z}} \boldsymbol{h} = c_i \partial_{\boldsymbol{z}} \boldsymbol{h} \; .$$

Finding the eigenvalues c_i and the corresponding eigenfunctions, the preserved function $\boldsymbol{h}(\boldsymbol{z})$ can be found by integrating the eigenfunctions. The function $\boldsymbol{h}(\boldsymbol{z})$ is called Riemann Invariant of the wave equation at hand (cf. [29] or [7] for more details). For the linear problem, introduced in Ex. 2.5 and analyzed in Ex. 4.1, the Riemann Invariants can be determined as follows.

The eigenvectors φ_i associated with the eigenvalues $c_i = (-1)^{i+1}\left(E\rho^{-1}\right)^{\frac{1}{2}}$ of the matrix

$$\left(\boldsymbol{D}^{-1}\boldsymbol{E}\right)^T = \begin{bmatrix} 0 & -EA \\ -(\rho A)^{-1} & 0 \end{bmatrix}$$

can be obtained as

$$\boldsymbol{\varphi}_i = (\partial_z h)_i = \begin{bmatrix} -\rho A c_i & 1 \end{bmatrix}^T . \tag{4.46}$$

Integrating these eigenvectors (4.46) yields the Riemann Invariants

$$h_i = n - \rho A c_i q = const. \tag{4.47}$$

See Ex. 4.23, depicting both preserved functions of the problem at hand.

Remark 4.11 (Hodograph Transformation). *Alternatively to Rem. 4.10, Riemann Invariants might be found by applying a hodograph transformation to the partial differential equation at hand and searching subsequently for characteristic lines. The hodograph transformations is given by $\boldsymbol{\mathcal{F}}(\boldsymbol{z}) : \mathbb{R}^d \mapsto \mathbb{R}^d$ such that $\boldsymbol{x} \mapsto \boldsymbol{\mathcal{F}}^{-1}(\boldsymbol{z})$ (cf. e.g. [29]). For instance, system* (4.6) *with* (4.14)*, can be transformed by inserting the Jacobian*

$$\partial_z \boldsymbol{x} = \begin{bmatrix} \partial_q s & \partial_n s \\ \partial_q t & \partial_n t \end{bmatrix}$$

along with its inverse

$$\partial_{\boldsymbol{x}} \boldsymbol{z} = \begin{bmatrix} \partial_s q & \partial_t q \\ \partial_s n & \partial_t n \end{bmatrix} = (\partial_z \boldsymbol{x})^{-1} = (\partial_q s \partial_n t - \partial_n s \partial_q t)^{-1} \begin{bmatrix} \partial_n t & -\partial_n s \\ -\partial_q t & \partial_q s \end{bmatrix}$$

into the system of first-order partial differential equations

$$\hat{\boldsymbol{D}}(\boldsymbol{z})\partial_q \boldsymbol{x} + \hat{\boldsymbol{E}}(\boldsymbol{z})\partial_n \boldsymbol{x} = \hat{\boldsymbol{F}}(\boldsymbol{z}) . \tag{4.48}$$

In Eq. (4.48) *the following coefficients have been introduced:*

$$\hat{\boldsymbol{D}} = \begin{bmatrix} 0 & 1 \\ 1 & 0 \end{bmatrix} \quad \hat{\boldsymbol{E}} = \begin{bmatrix} -\rho A_0 & 0 \\ 0 & -EA_0 \end{bmatrix} \quad \hat{\boldsymbol{F}} = \begin{bmatrix} 0 \\ 0 \end{bmatrix} . \tag{4.49}$$

Evaluating the directionality condition leads to the desired Riemann Invariants. A line $q = k(n)$ is a characteristic line if

$$\det\left(\hat{\boldsymbol{E}} - \hat{\boldsymbol{D}}\frac{dn}{dq}\right) = 0,$$

hence

$$(\mathrm{D}_q n)^2 = \rho A_0 E A_0 \qquad \Leftrightarrow \qquad \mathrm{D}_q n = \pm \rho A_0 \sqrt{\frac{E}{\rho}} \,. \tag{4.50}$$

By integrating Eq. (4.50)*, the Riemann Invariants are obtained as*

$$h_i = n - c_i \rho A_0 q = const. \tag{4.51}$$

Compare Eq. (4.51) *with Eq.* (4.47).

Remark 4.12 (Geometrical interpretation: The general non-linear case). *To gain more insights into hyperbolic problems in general, it might be beneficial to reflect some geometrical aspects of general first-order non-linear partial differential equations as given by*

$$F(s,t,z,p,q) = 0 \,. \tag{4.52}$$

Therein $p \equiv \partial_s z$ *and* $q \equiv \partial_t z$ *have been introduced denoting the partial derivatives of the solution* $z(s,t) : \Omega \mapsto \mathcal{Z} \subset \mathbb{R}$ *with respect to* $s \in \mathcal{S} \subset \mathbb{R}$ *and* $t \in \mathcal{T} \subset \mathbb{R}$ *where* $\Omega = \mathcal{S} \times \mathcal{T}$*. Solving Eq.* (4.52) *basically implies to find an integral surface* $z(s,t)$ *that satisfies Eq.* (4.52) *along with given boundary and initial conditions. In analogy to the basic concept of direction fields within the theory of ordinary differential equations (cf. [12] and [92]), the partial differential equation* (4.52) *allocates to every point* $P(s_P, t_P, z_P) \in \mathcal{A} = \mathcal{S} \times \mathcal{T} \times \mathcal{Z} \subset \mathbb{R}^3$ *planes*

$$(s_P - s)p + (t_P - t)q - (z_P - z) = 0, \tag{4.53}$$

that are tangential to integral surfaces of Eq. (4.52)*. Thus, Eq.* (4.52) *inheres the orientation of the tangent planes of the solution* $z(s,t)$ *at each point* $P \in \mathcal{A}$*. Since we claim that* $(\partial_p F)^2 + (\partial_q F)^2 \neq 0$*, Eq.* (4.52) *forms a one-parametric family for the partial derivatives* $\lambda \mapsto p(\lambda)$ *and* $\lambda \mapsto q(\lambda)$ *such that the planes* (4.53) *are envelopping surfaces of cones (see Fig. 4.11). To each point* $P \in \mathcal{A}$ *one cone is allocated, touching the integral surface. In the special case of (quasi-) linear partial differential equations these cones degenerate to single lines and the envelopping surfaces form a pencil of planes (cf. Rem. 4.13).*

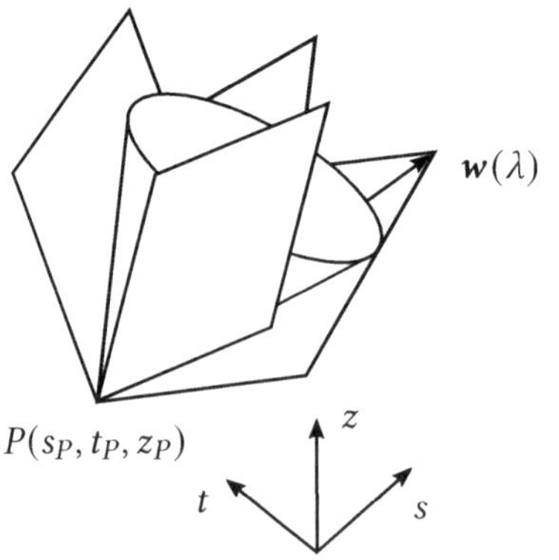

Figure 4.11.: Integral surfaces envelopping Monge's cone.

These cones are referred to as Monge's cones, and its generating lines are spanning a one-parametric vector field $\boldsymbol{w} : \mathbb{R} \mapsto \mathcal{A}$ *as given by*

$$\boldsymbol{w}(\lambda) = \begin{bmatrix} \partial_p F \\ \partial_q F \\ p\partial_p F + q\partial_q F \end{bmatrix} .$$

It is called Monge's vector field and curves $k(\xi)$ *whose tangents* $\mathrm{D}_\xi k(\xi)$ *belong to Monge's vector field, are referred to as focal curves or Monge's curves. This may be achieved by comparing the derivative of Eq.* (4.52) *with respect to* λ

$$\mathrm{D}_\lambda F(s,t,z,p(\lambda),q(\lambda)) = \partial_p F\,\mathrm{D}_\lambda p(\lambda) + \partial_q F\,\mathrm{D}_\lambda q(\lambda) = 0$$

as well as the derivative of the envelopping surfaces, introduced in (4.53), *both with respect to* λ

$$(s_P - s)\,\mathrm{D}_\lambda p(\lambda) + (t_P - t)\,\mathrm{D}_\lambda q(\lambda) = 0$$

such that the following relation of the plane increments can be established:

$$(s_P - s) : (t_P - t) : (z_P - z) = \partial_p F : \partial_q F : (p\partial_p F + q\partial_q F).$$

If $z(s,t)$ *is a solution of* (4.52), *then Monge's curves are obviously part of the integral surface. Consequently, they are called characteristic curves.*

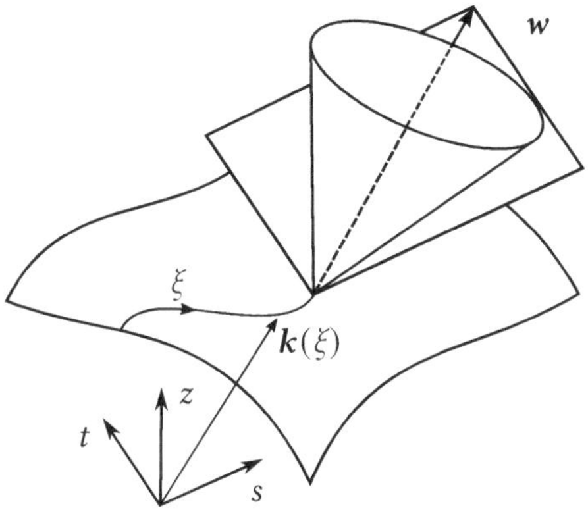

Figure 4.12.: Monge's cone with integral surface of initial boundary value problem along characteristic curve $k(\xi)$.

Along these characteristic lines, the partial differential equation (4.52) *can then equivalently be replaced by a system of ordinary differential equations as given by*

$$\mathrm{D}_\xi \boldsymbol{k}(\xi) = \mathrm{D}_\xi \begin{bmatrix} s \\ t \\ z \end{bmatrix} = \begin{bmatrix} \partial_p F \\ \partial_q F \\ p\partial_p F + q\partial_q F \end{bmatrix} = \boldsymbol{w} . \tag{4.54}$$

Alternatively, sytem (4.54) *might be reformulated as*

$$\mathrm{D}_t s = \frac{\partial_p F}{\partial_q F}, \qquad \mathrm{D}_t z = q + \frac{\partial_p F}{\partial_q F} p . \tag{4.55}$$

In addition to that, the set of all tangent planes along Monge's curves are referred to as Monge's strips. In case of $z(s,t)$ is a solution to the differential equation (4.52) *then these strips are called characteristic strips. These characteristic strips give rise to two further ordinary differential equations revealing the change of the partial derivatives q and p with respect to ξ*

$$\mathrm{D}_\xi p = -\partial_s F - p\partial_z F, \qquad \mathrm{D}_\xi q = -\partial_t F - q\partial_z F \tag{4.56}$$

or alternatively with respect to t

$$\mathrm{D}_t p = -\frac{\partial_s F}{\partial_q F} - p\frac{\partial_z F}{\partial_q F}, \qquad \mathrm{D}_t q = -\frac{\partial_t F}{\partial_q F} - q\frac{\partial_z F}{\partial_q F} . \tag{4.57}$$

This is obtained by analyzing the derivative of Eq. (4.52) *with respect to s*

$$\mathrm{D}_s F(s,t,z(s,t),p(s,t),q(s,t)) = \partial_s F + p\partial_z F + \partial_p F \partial_s p + \partial_q F \partial_s q = 0 .$$

By taking additionally advantage of the equality of mixed partials, i.e. $\partial^2_{st} z = \partial^2_{ts} z$ e.g. $\partial_s q = \partial_t p$, such that the following relation can be established

$$\partial_s F + p\partial_z F + \partial_p F \partial_s p + \partial_q F \partial_t p = 0 . \tag{4.58}$$

Then Monge's vector field (4.54) *gives rise to $\partial_p F = \mathrm{D}_\xi s$ and $\partial_q F = \mathrm{D}_\xi t$, hence Eq.* (4.58) *assumes the form*

$$\partial_s F + p\partial_z F + \mathrm{D}_\xi s \partial_s p + \mathrm{D}_\xi t \partial_t p = \partial_s F + p\partial_z F + \mathrm{D}_\xi p = 0 .$$

Herein use of $\mathrm{D}_\xi p = (\mathrm{D}_\xi s)(\partial_s p) + (\mathrm{D}_\xi t)(\partial_t p)$ has been made. Accordingly, by taking the derivative of Eq. (4.52) *with respect to the variable t, the following holds:*

$$\mathrm{D}_t F(s,t,z(s,t),p(s,t),q(s,t)) = \partial_t F + q\partial_z F + \mathrm{D}_\xi s \partial_s q + \mathrm{D}_\xi t \partial_t q = \partial_t F + q\partial_z F + \mathrm{D}_\xi q = 0 .$$

For more details on the geometry of partial differential equations, see [84, pp.41-49] , [31], [11, pp.369-370], [33] or [37].

Remark 4.13 (Geometrical interpretation: The special quasi-linear case)**.** *Based on the results proposed in Rem. 4.12, the special case of quasi-linear partial differential equations, assuming the form*

$$F(s,t,z,\partial_s z,\partial_t z) = a(s,t,z)\partial_t z + b(s,t,z)\partial_s z - c(s,t,z) = 0 , \tag{4.59}$$

is analyzed subsequently. Quasi-linearity implies linearity in the highest occuring partial derivatives, i.e. the coefficients a, b and c in (4.59) *are allowed to depend on the space and* *time variable as well as on the solution itself. Consequently, the coefficients are forming a vector field*

$$\mathbf{w}(s,t,z) = \begin{bmatrix} a & b & c \end{bmatrix}^T \tag{4.60}$$

that is perpendicular to the (normal) vector field

$$\boldsymbol{n} = \begin{bmatrix} \partial_t z & \partial_s z & -1 \end{bmatrix}^T .$$

Hence, $\mathbf{w}(s,t,z)$ *is tangential to the solution* $z(s,t)$ *in every point* $P \in \mathcal{A} = \mathcal{S} \times \mathcal{T} \times \mathcal{Z} \subset \mathbb{R}^3$*. The vector field* $\mathbf{w}(s,t,z)$ *is the Monge's vector field, introduced in Rem. 4.12. In each point* $P \in \mathcal{A}$ *Monge's vector field forms an axis of intersecting planes that are tangential to the integral surfaces. The orientation of Monge's vector field gives rise to a system of ordinary differential equations given by*

$$ds : dt : du = a : b : c . \tag{4.61}$$

These equations are equivalently governing the solution of (4.59) *along characteristic curves.*

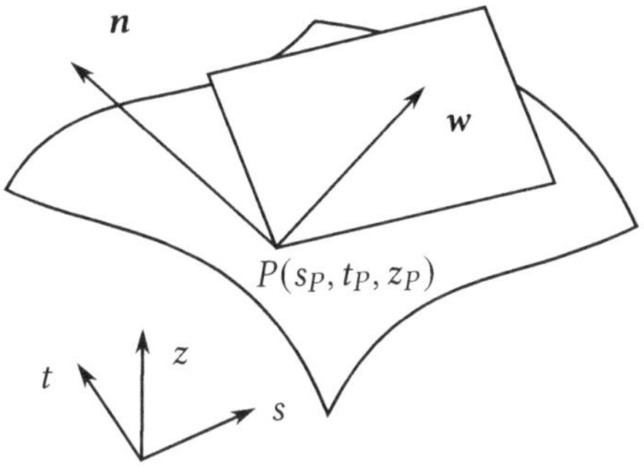

Figure 4.13.: Integral surface (contact element) $z - z_0 = p_0(s - s_0) + q_0(t - t_0)$ at point P_0 with normal and tangential vector field n and w, respectively.

Considering an initial line along which the solution is given and is everywhere tangential to Monge's vector field w *it becomes evident, that data which is considered as given only along such lines does not suffice to obtain an unique solution for the tangential planes. Hence, discontinuities of the partial derivatives* $\partial_s z$ *and* $\partial_t z$*, propagating along these characteristic lines, need to be permitted strictly. Consequently,*

$$\mathrm{d}z = \partial_s z\,\mathrm{d}s + \partial_t z\,\mathrm{d}t \qquad \text{i.e.} \qquad \partial_s z = z' - \partial_t z k'(s) \tag{4.62}$$

along with the given quasi-linear partial differential equation (4.59) *lead to an indifferent solution for the partial derivatives* $\partial_s z$ *and* $\partial_t z$*.*

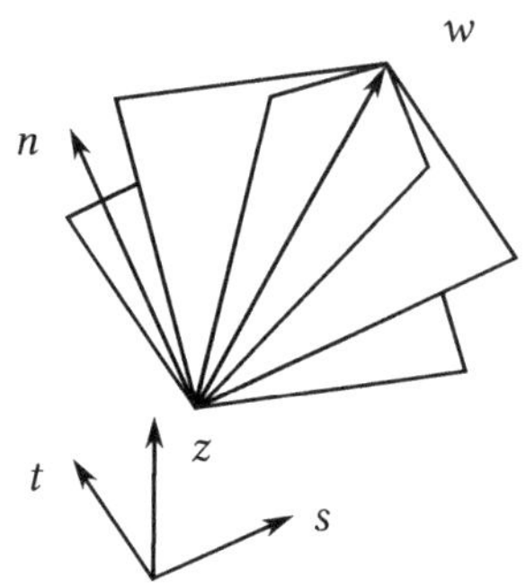

Figure 4.14.: Monge's pencil of planes.

Example 4.14 (Linear transport equation). *In the simplest case, linear transport phenomena are governed by partial differential equations in form of*

$$F(p, q) = ap + bq - c = 0\ , \tag{4.63}$$

where a, b and c are assumed to be constant. The partial derivatives of the solution z with respect to both independent variables s and t are defined as $p \equiv \partial_s z$ *and* $q \equiv \partial_t z$*, respectively. In accordance to Rem. 4.12 and Rem. 4.13, the differential equation* (4.63) *gives rise to a parameterization of the partial derivatives p and q, as given by*

$$q = \frac{c}{b}\lambda \quad \textit{and} \quad p = \frac{c}{a}(1 - \lambda)\ , \tag{4.64}$$

such that Monge's vector field

$$\mathbf{w} = \begin{bmatrix} \partial_p F \\ \partial_q F \\ p\partial_p F + q\partial_q F \end{bmatrix} = \begin{bmatrix} a \\ b \\ c \end{bmatrix}$$

can be identified as the axis of a pencil of planes given by

$$z - z_P = (s - s_P)p + (t - t_P)q = \frac{c}{a}(s - s_P)(1 - \lambda) + \frac{c}{b}(t - t_P)\lambda\ . \tag{4.65}$$

Thus, along characteristic lines, the solution is governed by a system of ordinary differential equations given by

$$\mathrm{D}_t s = \frac{\partial_p F}{\partial_q F} = \frac{a}{b} \quad \textit{and} \quad \mathrm{D}_t z \ = q + \frac{\partial_p F}{\partial_q F} p = q + \frac{a}{b} p = \frac{c}{b}\ .$$

Due to the linear structure of the underlying equation, the partial derivatives q and p are only changing linearly with respect to s and t, i.e. geometrically the characteristic stripes are not distorted. Consequently,

$$\mathrm{D}_t p = 0 \quad \textit{and} \quad \mathrm{D}_t q \ = 0\ .$$

Example 4.15 (Eikonal equation (cf. [37])). *Maybe the simplest representative of general non-linear first-order partial differential equations is the Eikonal equation given by*

$$F(s, t, z, p, q) = p^2 + q^2 - 1 = 0\ . \tag{4.66}$$

According to Rem. 4.12 and Rem. 4.13 the corresponding Monge's vecor field takes the form

$$\boldsymbol{w} = \begin{bmatrix} \partial_p F \\ \partial_q F \\ p\partial_p F + q\partial_q F \end{bmatrix} = 2 \begin{bmatrix} p \\ q \\ p^2 + q^2 \end{bmatrix} = 2 \begin{bmatrix} p \\ q \\ 1 \end{bmatrix}\ .$$

By introducing the parameterization

$$p = \cos\lambda \quad \textit{and} \quad q = \sin\lambda$$

a one-parametric vector field $w(\lambda) : \mathbb{R} \mapsto \mathcal{A}$ *can be obtained as*

$$\boldsymbol{w}(\lambda) = 2 \begin{bmatrix} \cos\lambda \\ \sin\lambda \\ 1 \end{bmatrix} \tag{4.67}$$

determining generating lines for Monge's cones. This cones are envelopped by the tangential planes

$$(z - z_P) = p(s - s_P) + q(t - t_P) = \cos\lambda(s - s_P) + \sin\lambda(t - t_P)\ .$$

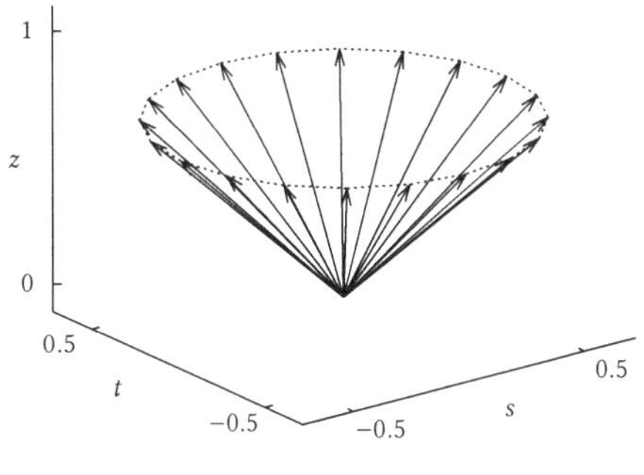

Figure 4.15.: Normalized Monge's vector field for the Eikonal equation (4.66).

Along characteristic lines Eq. (4.66) *can equivalently be transformed into a system of ordinary differential equations given by*

$$\mathrm{D}_t s = \frac{\partial_p F}{\partial_q F} = \frac{p}{q} \qquad \textit{and} \qquad \mathrm{D}_t z = q + \frac{\partial_p F}{\partial_q F} p = q + \frac{p^2}{q} = q\left(1 + \left(\frac{p}{q}\right)^2\right)\ .$$

Although, the Eikonal equation (4.66) is a non-linear equation, Monge's cones do not change its shape in the domain $\mathcal{A}$. Consequently, Monge's strips are not distorted, i.e.

$$\mathrm{D}_t p = 0 \qquad \textit{and} \qquad \mathrm{D}_t q = 0 \, .$$

Example 4.16 (Hamilton-Jacobi equations). *Another prominent representative within the class of general non-linear partial differential equations is given by*

$$F(r,t,S,p,q) = \partial_t S + H(r,t,\partial_r S) = q + H(r,t,p) = 0 \, . \tag{4.68}$$

This non-linear first-order partial differential equation, governing the action

$$S = \int L(t,r,\mathrm{D}_t r)\,\mathrm{d}t \, ,$$

is called the Hamilton-Jacobi equation. In Eq. (4.68) the function H denotes the Hamiltonian of the system and the partial derivatives with respect to time $t \in \mathcal{T}$ and configuration $r \in \mathbb{R}^{n\cdot d}$ are defined as

$$q \equiv \partial_t S \qquad \textit{and} \qquad p \equiv \partial_r S \, .$$

Following the geometrical interpretation of general non-linear partial differential equations, analyzed in Rem. 4.12, the Hamilton-Jacobi equation (4.68) can be transformed into a system of ordinary differential equations along characteristic lines. Evaluating (4.54)/(4.55) yields

$$\mathrm{D}_t r = \frac{\partial_p F}{\partial_q F} = \partial_p H \, , \tag{4.69}$$

$$\mathrm{D}_t S = q + p\frac{\partial_p F}{\partial_q F} = q + p\partial_p H \, . \tag{4.70}$$

Due to the non-linearity, the partial derivatives q and p are not changing linearly anymore. Hence, along characteristic strips, (4.56)/(4.57), leads additionally to a system of ordinary differential equations given by

$$\mathrm{D}_t p = -\frac{\partial_r F}{\partial_q F} - p\frac{\partial_S F}{\partial_q F} = -\partial_r H \, , \tag{4.71}$$

$$\mathrm{D}_t q = -\frac{\partial_t F}{\partial_q F} - q\frac{\partial_S F}{\partial_q F} = -\partial_t H \, . \tag{4.72}$$

The two ordinary differential equations $(4.69)_1$ and $(4.71)_1$ can be identified as the well-known canonical Hamiltonian equations. These equations are the characteristic directions of the Hamilton-Jacobi equation. Note that the action S does not occur explicitly in (4.68), such that $\partial_S F = 0$. Furthermore $\partial_q F = 1$ applies. For more details on the Hamilton-Jacobi theory, see e.g. [11, pp. 248-270], and [84, pp. 66-68].

Example. *Assuming natural, autonomous mechanical systems with Hamiltonian as given by*

$$H = p^2 + V(r) \, . \tag{4.73}$$

Then, the Hamilton-Jacobi equation (4.68) *is obtained as*

$$F(r,t,S,p,q) = q + p^2 + V(r) = 0 \ .$$

In accordance to Rem. 4.12 and Rem. 4.13, Monge's vector field is determined as

$$\mathbf{w} = \begin{bmatrix} \partial_p F \\ \partial_q F \\ p\partial_p F + q\partial_q F \end{bmatrix} = \begin{bmatrix} 2p \\ 1 \\ 2p^2 + q \end{bmatrix} \ . \tag{4.74}$$

By introducing the parameterisation

$$p = \lambda \qquad \text{and} \qquad q = -\lambda^2 - V(r)$$

Monge's vector field constitutes a one-parametric vector field

$$\mathbf{w}(\lambda) = \begin{bmatrix} 2\lambda \\ 1 \\ \lambda^2 - V(r) \end{bmatrix} \tag{4.75}$$

that gives rise to the generating lines of Monge's cones envelopped by the tangential planes

$$(S - S_0) = \lambda(r - r_0) - (\lambda^2 + V(r))(t - t_0).$$

Along characteristic lines the Hamilton-Jacobi equation (4.68) *can then equivalently be transformed into a system of ordinary differential equations as given by*

$$\mathrm{D}_t r = \frac{\partial_p F}{\partial_q F} = 2p \qquad \text{and} \qquad \mathrm{D}_t S = q + \frac{\partial_p F}{\partial_q F} p = q + 2p^2 \ .$$

In contrast to Ex. 4.14 and Ex. 4.15 for the Hamilton-Jacobi equations Monge's cones change its shape in any point $\mathcal{A}$, consequently Monge's strips are distorted. This leads to an additional system of ordinary differential equations given by

$$\mathrm{D}_t p = -\frac{\partial_r F}{\partial_q F} = -\mathrm{D}_r V(r) \qquad \text{and} \qquad \mathrm{D}_t q = -\frac{\partial_t F}{\partial_q F} = 0 \ .$$

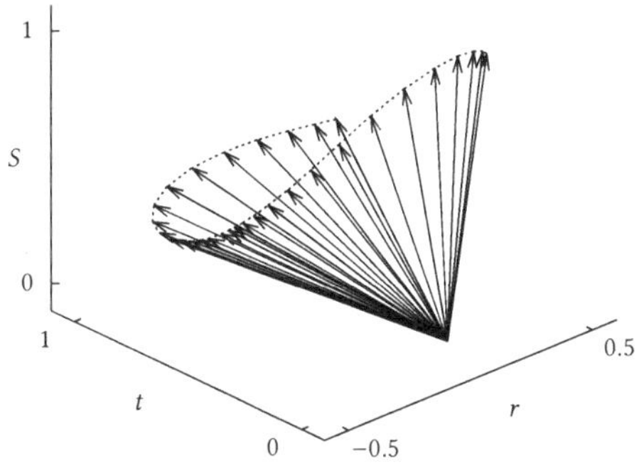

Figure 4.16.: Normalized Monge's vector field for the Hamilton-Jacobi equation (4.68) assuming the Hamiltonian (4.73).

Remark 4.17 (Plasticity). *The elasto-plastic behaviour of general continua is governed by quasi-linear hyperbolic partial differential equations as well (cf. e.g. [84, pp. 117-118] and [1]). Following [83] and [97], the characteristic lines can be identified coinciding with the directions of plastic flow.*

4.2. Space-time finite element method[3]

In Sec. 4.1 the hyperbolic structure of the governing equations has been revealed and consequences regarding the mechanisms of how information flows in physical time-dependent systems have been elaborated. The method of characteristics, proposed in Sec. 4.1 as well, relies on an integration of the partial differential equation simultaneously in space and time. This intuitively motivates the development of further integration methods, that are based on a simultaneous discretization of the space-time domain. Therfore, the present section aims to present novel Galerkin space-time finite element methods for the inverse dynamics of infinite-dimensional systems. For further informations on space-time finite element methods in general we would like to refer to the two pioneering publications [9] and [52]. See also [80], [6], [54], [53] and [47].

By introducing the velocity $\boldsymbol{v}(s,t) = \partial_t \boldsymbol{x}(s,t)$, the underlying partial differential equation at hand (1.1) can be transformed into a system of partial differential equations that is first-order in time:

$$\begin{aligned} \partial_t \boldsymbol{x} - \boldsymbol{v} &= \boldsymbol{0} \,, \\ \boldsymbol{A}\partial_t \boldsymbol{v} - \partial_s(\boldsymbol{B}\partial_s \boldsymbol{x}) &= \boldsymbol{C} \,. \end{aligned} \tag{4.76}$$

Multiplying the equations in (4.76) by sufficiently smooth test functions $\boldsymbol{w}_1, \boldsymbol{w}_2 : \Omega \mapsto \mathbb{R}^d$ and subsequently integrating over the space-time domain $\Omega = S \times \mathcal{T}$ yields

$$\begin{aligned} \int_\Omega \boldsymbol{w}_1 \cdot (\partial_t \boldsymbol{x} - \boldsymbol{v}) \, \mathrm{d}\Omega &= 0 \\ \int_\Omega \boldsymbol{w}_2 \cdot (\boldsymbol{A}\partial_t \boldsymbol{v} - \partial_s (\boldsymbol{B}\partial_s \boldsymbol{x})) \, \mathrm{d}\Omega &= \int_\Omega \boldsymbol{w}_2 \cdot \boldsymbol{C} \, \mathrm{d}\Omega \end{aligned} \tag{4.77}$$

Integrating by parts the second term on the left-hand side of $(4.77)_2$ and taking into account the boundary conditions $(2.11)_2$ yields

$$\begin{aligned} \int_\Omega \boldsymbol{w}_2 \cdot \boldsymbol{A}\partial_t \boldsymbol{v} \, \mathrm{d}\Omega + \int_\Omega \partial_s \boldsymbol{w}_2 \cdot \boldsymbol{B}\partial_s \boldsymbol{x} \, \mathrm{d}\Omega + \int_{\partial\Omega_f} \boldsymbol{f}(t) \cdot \boldsymbol{w}_2|_{(s,t)\in\partial\Omega_f} \, \mathrm{d}t \\ = \int_\Omega \boldsymbol{w}_2 \cdot \boldsymbol{C} \, \mathrm{d}\Omega + \int_{\partial\Omega_\eta} \boldsymbol{\eta}(t) \cdot \boldsymbol{w}_2|_{(s,t)\in\partial\Omega_\eta} \, \mathrm{d}t \,. \end{aligned} \tag{4.78}$$

[3] This section partly reproduces [93].

Concerning the prescribed boundary condition on $\partial\Omega_\eta$, we make use of weakly enforced servo-constraints of the form

$$\int_{\partial\Omega_\eta} \boldsymbol{w}_3(t) \cdot \left(\boldsymbol{\gamma}(t) - \boldsymbol{x}|_{(s,t)\in\partial\Omega_\eta}\right) dt = 0 \tag{4.79}$$

where $\boldsymbol{w}_3 : \partial\Omega_L \mapsto \mathbb{R}^d$ is a third test function.

Now the weak formulation of the inverse dynamics problem at hand can be stated as follows: Given $\boldsymbol{b}$, $\boldsymbol{f}_0$, $\boldsymbol{x}_0$ and $\boldsymbol{v}_0$, find $\boldsymbol{x} \in V_1$, $\boldsymbol{v} \in V_2$ and $\boldsymbol{f} \in V_3$, such that

$$\begin{aligned}
&\int_\Omega \boldsymbol{w}_1 \cdot \boldsymbol{A}(\partial_t \boldsymbol{x} - \boldsymbol{v})\, \mathrm{d}\Omega = 0 \\
&\int_\Omega \boldsymbol{w}_2 \cdot \boldsymbol{A}\partial_t \boldsymbol{v}\, \mathrm{d}\Omega + \int_\Omega \partial_s \boldsymbol{w}_2 \cdot \boldsymbol{B}\partial_s \boldsymbol{x}\, \mathrm{d}\Omega + \int_{\partial\Omega_f} \boldsymbol{f}(t) \cdot \boldsymbol{w}_2|_{(s,t)\in\partial\Omega_f}\, \mathrm{d}t \\
&\qquad = \int_\Omega \boldsymbol{w}_2 \cdot \boldsymbol{C}\, \mathrm{d}\Omega + \int_{\partial\Omega_\eta} \boldsymbol{\eta}(t) \cdot \boldsymbol{w}_2|_{(s,t)\in\partial\Omega_\eta}\, \mathrm{d}t \\
&\int_{\partial\Omega_\eta} \boldsymbol{w}_3(t) \cdot (\boldsymbol{\gamma}(t) - \boldsymbol{x}|_{(s,t)\in\partial\Omega_\eta})\, \mathrm{d}t = 0
\end{aligned} \tag{4.80}$$

for arbitrary test functions $\boldsymbol{w}_\alpha \in W_\alpha$ for $\alpha \in \{1,2\}$ and $\boldsymbol{w}_3 \in W_3$. The prescribed initial conditions (2.30) give rise to the Dirichlet-type boundary $\partial\Omega_0$ of the space-time domain Ω. Accordingly, we have

$$\begin{aligned}
V_1 &= \left\{\boldsymbol{x} : \Omega \mapsto \mathbb{R}^d \,|\, \boldsymbol{x} = \boldsymbol{x}_0 \text{ on } \partial\Omega_0\right\} \\
V_2 &= \left\{\boldsymbol{v} : \Omega \mapsto \mathbb{R}^d \,|\, \boldsymbol{v} = \boldsymbol{v}_0 \text{ on } \partial\Omega_0\right\} \\
V_3 &= \left\{\boldsymbol{f} : \partial\Omega_f \mapsto \mathbb{R}^d \,|\, \boldsymbol{f} = \boldsymbol{f}_0 \text{ on } \partial\Omega_f \cap \partial\Omega_0\right\}
\end{aligned}$$

and

$$\begin{aligned}
W_\alpha &= \left\{\boldsymbol{w}_\alpha : \Omega \mapsto \mathbb{R}^d \,|\, \boldsymbol{w}_\alpha = \boldsymbol{0} \text{ on } \partial\Omega_0\right\} \\
W_3 &= \left\{\boldsymbol{w}_3 : \partial\Omega_\eta \mapsto \mathbb{R}^d \,|\, \boldsymbol{w}_3 = \boldsymbol{0} \text{ on } \partial\Omega_\eta \cap \partial\Omega_0\right\}
\end{aligned}$$

for $\alpha \in \{1,2\}$. The space-time finite element discretization of the weak form (4.80) is based on finite-dimensional sub-spaces $V_i^h \subset V_i$ and $W_i^h \subset W_i$ associated with interpolations of the trial functions

$$\begin{aligned}
\boldsymbol{x}^h(s,t) &= \sum_{a=1}^{n_\Omega} N_a(s,t)\boldsymbol{x}_a \\
\boldsymbol{v}^h(s,t) &= \sum_{a=1}^{n_\Omega} N_a(s,t)\boldsymbol{v}_a
\end{aligned}
\qquad \text{and} \qquad
\begin{aligned}
\boldsymbol{f}^h(t) &= \sum_{i\in\mathcal{N}_0} L_i(t)\boldsymbol{f}_i \\
\boldsymbol{\eta}^h(t) &= \sum_{i\in\mathcal{N}_L} L_i(t)\boldsymbol{\eta}_i
\end{aligned}$$

where n_Ω is the total number of nodes giving rise to the index set $\mathcal{N}_\Omega = \{1,\ldots,n_\Omega\}$. Moreover, index set $\mathcal{N}_f \subset \mathcal{N}_\Omega$ contains the nodes lying on $\partial\Omega_f$. Similarly, $\mathcal{N}_\eta \subset \mathcal{N}_\Omega$ contains the nodes lying on $\partial\Omega_\eta$. The test functions are approximated by

$$\boldsymbol{w}_\alpha^h(s,t) = \sum_{a=1}^{n_\Omega} N_a(s,t)\boldsymbol{w}_{\alpha_a} \qquad \text{and} \qquad \boldsymbol{w}_3^h(t) = \sum_{i\in\mathcal{N}_\eta} L_i(t)\boldsymbol{w}_{3_i}$$

for $\alpha \in \{1, 2\}$.

Remark 4.18. *(Direct dynamics) The space-time finite element method has been introduced in Sec. 4.2 for the inverse dynamics problem. Therefore, the servo-constraint has been imposed weakly. In principle, ideal-orthogonal (contact-)constraints can be imposed in the same manner, such that the proposed space-time finite element method is also capable of solving the direct dynamics of constrained mechanical systems.*

Resulting algebraic formulation.

In the present work, we confine our attention to standard low-order isoparametric finite elements. In particular, L_i denote α-linear Lagrangian shape functions also employed in Chap. 3, while N_a denote $(\alpha + 1)$-linear Lagrangian shape functions. Inserting the above finite element approximations into weak form (4.80) yields the algebraic system of equations

$$\begin{aligned} \mathcal{A}Q - \mathcal{M}V &= \mathbf{0} \, , \\ \mathcal{A}V + \mathcal{F}^{\text{int}}(Q) + \mathcal{B}_f^T \mathcal{F}_f &= \mathcal{F}^{\text{ext}} + \mathcal{B}_\eta^T \mathcal{F}_\eta \, , \\ \mathcal{B}_\eta Q - \Gamma &= \mathbf{0} \end{aligned} \tag{4.81}$$

where

$$\begin{aligned} \mathcal{A}_{ab} &= I_d \int_\Omega A N_a \partial_t N_b \, \mathrm{d}\Omega \, , \\ \mathcal{M}_{ab} &= I_d \int_\Omega A N_a N_b \, \mathrm{d}\Omega \, , \\ \mathcal{F}_a^{\text{int}} &= \int_\Omega \partial_s N_a B \partial_s x^h \, \mathrm{d}\Omega \, , \\ \mathcal{B}_{f_{ib}} &= I_d \int_{\partial\Omega_f} L_i N_b \, \mathrm{d}t \, , \\ \mathcal{B}_{\eta_{ib}} &= I_d \int_{\partial\Omega_\eta} L_i N_b \, \mathrm{d}t \, , \\ \mathcal{F}_a^{\text{ext}} &= \int_\Omega N_a C \, \mathrm{d}\Omega \, , \\ \Gamma_i &= \int_{\partial\Omega_\eta} L_i \gamma \, \mathrm{d}t \, . \end{aligned} \tag{4.82}$$

In (4.81), the unknown nodal quantities x_a, v_a and f_i have been assembled in corresponding nodal system vectors Q, V and $\mathcal{F}_f$. Elimination of the nodal velocity vector V yields the residual vector

$$\mathcal{R}(Q, \mathcal{F}_f) = \begin{bmatrix} \mathcal{A}\mathcal{M}^{-1}\mathcal{A}Q + \mathcal{F}^{\text{int}}(Q) + \mathcal{B}_f^T \mathcal{F}_f - \mathcal{F}^{\text{ext}} - \mathcal{B}_\eta^T \mathcal{F}_\eta \\ \mathcal{B}_\eta Q - \Gamma \end{bmatrix} \tag{4.83}$$

The resulting algebraic system of non-linear equations, $\mathcal{R}(Q, \mathcal{F}_f) = \mathbf{0}$, can be solved iteratively for the nodal unknowns $(Q, \mathcal{F}_f)$ by applying Newton's method. In this connection, the iteration matrix is given by

$$D\mathcal{R}(Q, \mathcal{F}_f) = \begin{bmatrix} \mathcal{A}\mathcal{M}^{-1}\mathcal{A} + D\mathcal{F}^{\text{int}}(Q) & \mathcal{B}_f^T \\ \mathcal{B}_\eta & \mathbf{0} \end{bmatrix} \tag{4.84}$$

where $D\mathcal{F}^{\text{int}}(Q)$ results from $(4.82)_3$ together with the constitutive term $(2.12)_2$. A straightforward calculation yields the nodal contribution to $D\mathcal{F}^{\text{int}}(Q)$ given by

$$\partial_{x_b}\mathcal{F}_a^{\text{int}}(Q) = \int_\Omega \partial_s N_a \boldsymbol{H} \partial_s N_b \, \mathrm{d}\Omega \tag{4.85}$$

where matrix $\boldsymbol{H}$ has been defined in (4.5). Note that $\boldsymbol{p}$ therein has now to be substituted with $\boldsymbol{x}^h$.

Recursive implementation.

The newly devised space-time finite element method yields the system of algebraic equations (4.81) which can be solved by applying the iterative procedure outlined in Sec. 4.2. This implies the simultaneous solution of all the unknowns resulting from the space-time discretization.

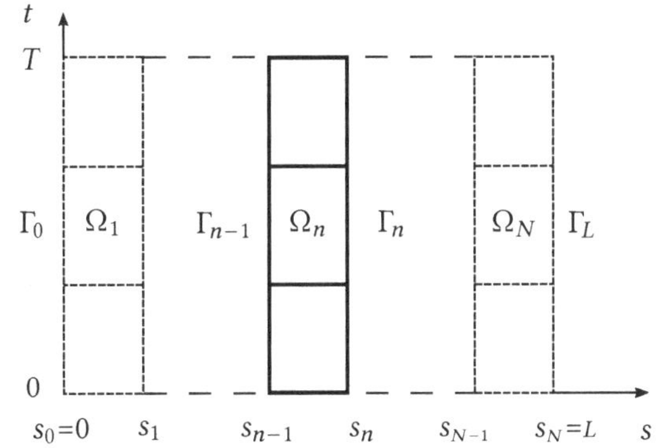

Figure 4.17.: Division of the space-time domain $\Omega = S \times T$ into time-space slabs $\Omega_n = (s_{n-1}, s_n) \times \mathcal{T}$ for $n = 1, \ldots, N$.

We next show that the solution can alternatively be attained by applying a recursive scheme which relies on the decomposition of the space-time domain Ω into N time-space slabs $\Omega_n = (s_{n-1}, s_n) \times \mathcal{T}$, $n = 1, \dots, N$ (cf. Fig.4.17). Focusing on one representative slab Ω_n, the recursive solution procedure relies on the weak form

$$
\begin{aligned}
\int_{\Omega_n} \boldsymbol{w}_1 \cdot \boldsymbol{A} \cdot (\partial_t \boldsymbol{x} - \boldsymbol{v}) \, \mathrm{d}\Omega &= 0 \\
\int_{\Omega_n} \boldsymbol{w}_2 \cdot \boldsymbol{A} \cdot \partial_t \boldsymbol{v} \, \mathrm{d}\Omega + \int_{\Omega_n} \partial_s \boldsymbol{w}_2 \cdot \boldsymbol{B} \partial_s \boldsymbol{x} \, \mathrm{d}\Omega + \int_{\Gamma_{n-1}} \boldsymbol{w}_2 \cdot \boldsymbol{f}_{n-1} \, \mathrm{d}t & \\
= \int_{\Omega_n} \boldsymbol{w}_2 \cdot \boldsymbol{C} \, \mathrm{d}\Omega + \int_{\Gamma_n} \boldsymbol{w}_2 \cdot \boldsymbol{f}_n \, \mathrm{d}t & \\
\int_{\Gamma_n} \boldsymbol{w}_{3_n} \cdot (\boldsymbol{x} - \boldsymbol{\gamma}_n) \, \mathrm{d}t &= 0 \\
\int_{\Gamma_{n-1}} \boldsymbol{w}_{3_{n-1}} \cdot (\boldsymbol{x} - \boldsymbol{\gamma}_{n-1}) \, \mathrm{d}t &= 0
\end{aligned}
\tag{4.86}
$$

where $\Gamma_n = \{s_n\} \times \mathcal{T}$. Note that $\Gamma_N = \partial\Omega_\eta$ and $\Gamma_0 = \partial\Omega_f$ is defined. In essence, weak form (4.86) emanates from the original formulation (4.80) by restricting the space-time domain to Ω_n. In this connection, $(4.86)_{3,4}$ have been introduced to facilitate a convenient description of the recursive solution procedure. The overall space-time approximation remains the same as before. The approximation of the newly introduced intermediate quantities $\boldsymbol{f}_n$ $(n = 1, \dots, N - 1)$ is given by

$$
\boldsymbol{f}_n^h(t) = \sum_{i \in \mathcal{N}_n} L_i(t) \boldsymbol{f}_{n_i} \tag{4.87}
$$

where index set $\mathcal{N}_n$ contains the nodes lying on Γ_n. In this way, $\boldsymbol{f}_n^h \in V_{3_n}^h \subset V_{3_n}$, where

$$
V_{3_n} = \left\{ \boldsymbol{f}_n : \Gamma_n \mapsto \mathbb{R}^d \mid \boldsymbol{f}_n = \bar{\boldsymbol{f}}_n \text{ on } \Gamma_n \cap \partial\Omega_0 \right\}
$$

and $\bar{\boldsymbol{f}}_n$ $(n = 1, \dots, N - 1)$ have to satisfiy the initial conditions. Similarly, we introduce

$$
W_{3_n} = \left\{ \boldsymbol{w}_{3_n} : \Gamma_n \mapsto \mathbb{R}^d \mid \boldsymbol{w}_{3_n} = \boldsymbol{0} \text{ on } \Gamma_n \cap \partial\Omega_0 \right\}
$$

for $n = 0, \dots, N - 1$, so that in $(4.86)_{3,4}$, $\boldsymbol{w}_{3_n} \in W_{3_n}$. Correspondingly, $\boldsymbol{w}_{3_n}^h \in W_{3_n}^h \subset W_{3_n}$ is given by

$$
\boldsymbol{w}_{3_n}^h = \sum_{i \in \mathcal{N}_n} L_i(t) (\boldsymbol{w}_{3_n})_i
$$

It can be easily shown that the recursive application of (4.86) for $n = 1, \dots, N$ is equivalent to the original space-time method, provided that the approximation of $\boldsymbol{\gamma}_n$ in $(4.86)_{3,4}$ conforms with the continuity of the displacement approximation. To ensure this, we choose

$$
\boldsymbol{\gamma}_n^h(t) = \sum_{i \in \mathcal{N}_n} L_i(t) \boldsymbol{\gamma}_{n_i} \tag{4.88}
$$

so that $(4.86)_{3,4}$ enforce $r^h|_{\Gamma_n} = \gamma_n^h$ and $r^h|_{\Gamma_{n-1}} = \gamma_{n-1}^h$. Now, the equivalence between $(4.86)_1$ and $(4.80)_1$ follows from the property

$$\sum_{n=1}^{N} \int_{\Omega_n} (\dots)\, \mathrm{d}\Omega = \int_{\Omega} (\dots)\, \mathrm{d}\Omega\,.$$

Similarly, the equivalence between Eq. $(4.86)_2$ and Eq. $(4.80)_2$ follows by additionally taking into account

$$\sum_{n=1}^{N} \left(\int_{\Gamma_{n-1}} w_2 \cdot f_{n-1}\, \mathrm{d}t - \int_{\Gamma_n} w_2 \cdot f_n\, \mathrm{d}t \right) = \int_{\Gamma_0} w_2 \cdot f_0\, \mathrm{d}t - \int_{\Gamma_L} w_2 \cdot f_L\, \mathrm{d}t\,.$$

Moreover, $(4.86)_3$ contains $(4.80)_3$ for $n = N$ under the provision that the prescribed trajectory γ on Γ_L is based on nodal interpolation of the type (4.88).

Now the proposed space-stepping algorithm results from the recursive application of the discretized weak form (4.86) in backward space direction, that is, for $n = N, N-1, \dots, 1$ as shown in Tab. 4.1.

LOAD γ_N^h, f_N^h
DO n=N,$\cdots$,1
SOLVE Eq.(4.86) $\longrightarrow$ $r^h\|_{\Omega_n}$, $v^h\|_{\Omega_n}$, γ_{n-1}^h, f_{n-1}^h

Table 4.1.: Recursive implementation.

Note that for $n = N$ the given data follows from the prescribed data on Γ_L leading to $\gamma_N^h = \gamma^h$ and $f_N^h = f_L^h$, where the prescribed data γ and f_L is assumed to be given in interpolated form of the type (4.88) and (4.87), respectively.

The implementation of the recursive scheme is again based on the algebraic formulation outlined in Sec. 4.2. However, in each step of the recursive scheme all of the algebraic quantities are now confined to the respective slab Ω_n (see also Rem. 4.20).

Remark 4.19. *The meaning of the newly introduced intermediate quantities f_n and f_{n-1} ($n = 2, \dots, N-1$) in Eq.* $(4.86)_2$ *can be elucidated by considering the Euler-Lagrange equations emanating from Eq.* $(4.86)_2$. *A straightforward calculation based on integration by parts applied to the second term on the left-hand side of Eq.* $(4.86)_2$ *yields*

$$\int_{\Omega_n} w_2 \cdot (A\partial_t v - \partial_s(B\partial_s x) - C)\, \mathrm{d}\Omega + \int_{\Gamma_n} w_2 \cdot (n - f_n)\, \mathrm{d}t + \int_{\Gamma_{n-1}} w_2 \cdot (f_{n-1} - n)\, \mathrm{d}t = 0$$

Accordingly, $f_n = n|_{\Gamma_n}$ and $f_{n-1} = n|_{\Gamma_{n-1}}$, so that f_n and f_{n-1} correspond to the contact forces in the string along Γ_n and Γ_{n-1}, respectively.

Remark 4.20 (Numerical effort). *Assuming a regular discretization of the space-time domain $\Omega = S \times T$ relying on n_s nodes in space direction and n_t nodes in time direction amounts to a total number of nodes $n_\Omega = n_s \cdot n_t$. The simultaneous solution for all unknowns outlined in Sec. 4.2 relies on the iteration matrix* (4.84)*, whose dimension is equal to $(n_s+1)\cdot(n_t-1)\cdot d$. In contrast to that, the dimension of the iteration matrix corresponding to each of the $N = n_s - 1$ steps of the recursive solution procedure is equal to $(2\alpha+1)\cdot(n_t-1)\cdot d$.*

Remark 4.21. *The semi-discrete approach described in Sec. 3.1 can be linked to the space-time finite element method developed above. To this end, we consider a Galerkin-based discretization in time of the semi-discrete formulation. In particular, we introduce the following approximations of the nodal position and velocity vectors*

$$\begin{aligned} \boldsymbol{q}_i^h(t) &= L_j(t)\boldsymbol{q}_{ij} \\ \boldsymbol{v}_i^h(t) &= L_j(t)\boldsymbol{v}_{ij} \end{aligned} \tag{4.89}$$

For conciseness the summation convention applies throughout this remark. Moreover, for simplicity we again employ Lagrangian shape functions in (4.89)*. In addition to that, we introduce nodal weighting functions of the form*

$$\boldsymbol{W}_i^h(t) = M_A(t)\boldsymbol{W}_{iA} \tag{4.90}$$

where M_A are basis functions to be specified below. Now the semi-discrete equations (3.51) *can be recast in the weighted Galerkin form*

$$\begin{aligned} \int_{\mathcal{T}} \boldsymbol{W}_i^h \cdot \boldsymbol{M}_{ij}\left(\mathrm{D}_t\boldsymbol{q}_j^h - \boldsymbol{v}_j^h\right)\,\mathrm{d}t &= 0 \\ \int_{\mathcal{T}} \boldsymbol{W}_i^h \cdot \left(\boldsymbol{M}_{ij}\,\mathrm{D}_t\boldsymbol{v}_j^h + \boldsymbol{f}_i^{\mathrm{int}}(\boldsymbol{q}^h) - \boldsymbol{f}_i^{\mathrm{ext}}(t) + \delta_{i1}\boldsymbol{f}_0^h(t) - \delta_{i,p+1}\boldsymbol{f}_L^h(t)\right)\,\mathrm{d}t &= 0 \\ \int_{\mathcal{T}} \boldsymbol{W}_i^h \cdot \delta_{i,p+1}\left(\boldsymbol{q}_i^h - \boldsymbol{\gamma}\right)\,\mathrm{d}t &= 0 \end{aligned} \tag{4.91}$$

Note that $(4.91)_1$ *serves the purpose of introducing the nodal velocities $\boldsymbol{v}_i^h$. Time-stepping schemes typically applied in the context of the semi-discrete formulation can be recovered from* (4.91) *by choosing specific* discontinuous *shape functions M_A in* (4.90)*, along with associated quadrature formulas for the evaluation of the time integrals. This approach gives rise to the so-called continuous Galerkin method*[4] *(see [36]). For example, focusing on the first two equations in* (4.91) *emanating from the underlying system of first-order ordinary differential equations, choosing constant shape functions M_A and linear Lagrangian shape functions L_j, along with the mid-point quadrature for the evaluation of the time integrals yields the mid-point rule (cf. e.g. [19]).*

[4] Note that the designation *continuous Galerkin* is attributed to continuous test functions despite the use of discontinuous weighting functions. This is in contrast to the so-called discontinuous Galerkin method which relies on test and weighting functions, both being discontinuous (cf. [36]).

In sharp contrast to that, the newly developed space-time finite element method can be recovered from (4.91) *by choosing* continuous *shape functions* M_A. *In particular, using again Lagrangian shape functions, i.e.* $M_A(t) = L_A(t)$, (4.91) *leads to the algebraic formulation* (4.81), *provided that*

(i) the shape functions $N_a(s, t)$ *used in the above space-time finite element method have tensor product form (cf. [16, 81], see also [99]). That is,*

$$N_a(\mathbf{s}, t) = L_i(t) L_j(s_1) L_k(s_2) \cdot L_m(s_\alpha) \; . \tag{4.92}$$

Referring to the local node numbering of a α*-linear master element the node numbers in* (4.92) *are related to each other. In Fig. 4.18 and Tab. 4.2, this relation is illustrated exemplarily for* $\alpha = 1$.

a	i	j
1	1	1
2	2	1
3	2	2
4	1	2

Table 4.2.: Relation between the local node numbers of the shape functions in (4.92).

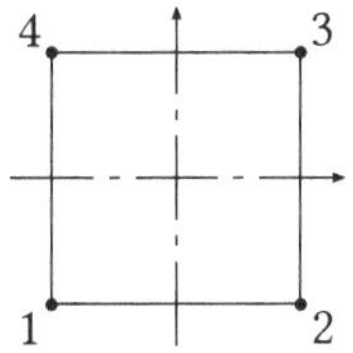

Figure 4.18.: Bi-linear master element with local node numbering.

In this connection, the one-dimensional linear Lagrangian shape functions are given by

$$\begin{aligned} L_1(\xi) &= \frac{\xi_2 - \xi}{\xi_2 - \xi_1} \\ L_2(\xi) &= \frac{\xi - \xi_1}{\xi_2 - \xi_1} \end{aligned} \tag{4.93}$$

where ξ_1 *and* ξ_2 *refer to the nodal coordinates, while* ξ *stands for either* $s \in \mathcal{S}$ *or* $t \in \mathcal{T}$.

(ii) rectangular shaped space-time finite elements are used leading to a structured mesh in space-time. Note that this precludes unstructured space-time meshes which are feasible with the space-time finite element method proposed above.

To further illustrate the connection between (4.91) *and* (4.81), *we consider the first term in* $(4.91)_2$, *which yields*

$$\begin{aligned} \int_{\mathcal{T}} \mathbf{W}_i^h \cdot \mathbf{M}_{ij} \, \mathrm{D}_t \boldsymbol{v}_j^h \, \mathrm{d}t &= \mathbf{W}_{iA} \cdot \int_{\mathcal{T}} L_A(t) L_k'(t) \, \mathrm{d}t \, \mathbf{M}_{ij} \boldsymbol{v}_{jk} \\ &= \mathbf{W}_{iA} \cdot \int_{\mathcal{T}} L_A(t) L_k'(t) \, \mathrm{d}t \int_{\mathcal{S}} A_\rho L_i(s) L_j(s) \, \mathrm{d}s \, \boldsymbol{v}_{jk} \\ &= \underbrace{\mathbf{W}_{iA}}_{\mathbf{w}_{2_a}} \cdot \int_{\Omega} A_\rho \underbrace{L_i(s) L_A(t)}_{N_a(s,t)} \underbrace{L_j(s) L_k'(t)}_{\partial_t N_b(s,t)} \, \mathrm{d}\Omega \; \underbrace{\boldsymbol{v}_{jk}}_{\mathbf{V}_b} \\ &= \mathbf{w}_{2_a} \cdot \mathcal{A}_{ab} \mathbf{V}_b \end{aligned}$$

Here, $\boldsymbol{M}_{ij}$ introduced in (3.7) has been used. Moreover, the relation between the nodal shape functions in Tab. 4.2 has been taken into account.

The remaining terms in (4.91) can be linked to the corresponding terms in (4.81) in a similar manner.

Remark 4.22. *[A discontinuous Galerkin approach] Due to the hyperbolic structure, the solution of the partial differential equation is dominated by wave phenomena. The mechanisms of wave propagation have lead to the perception that a simultaneous space-time integration might be the natural choice of solving the inverse dynamics problem of infinite-dimensional systems. Therefore a modified discontinuous Galerkin method is derived below. Its virtue lies in not being confined to spatially continuous systems. We are considering ordinary differential equations in form of*

$$\mathrm{D}_t \boldsymbol{z} + \boldsymbol{\mathcal{F}}(\boldsymbol{z}, t) + \boldsymbol{\mathcal{U}} = \boldsymbol{0} \tag{4.94}$$

governing the solution $\boldsymbol{z} = \begin{bmatrix} \boldsymbol{q} & \boldsymbol{v} \end{bmatrix}^T$ that is additionally subjected to servo-constraints as given by

$$\boldsymbol{h}(\boldsymbol{q}, t) = \boldsymbol{H}\boldsymbol{q} - \boldsymbol{y}(t) = \boldsymbol{0} \; . \tag{4.95}$$

In Eq. (4.94)

$$\boldsymbol{\mathcal{U}} = \begin{bmatrix} \boldsymbol{0} \\ \boldsymbol{B}_f \boldsymbol{f}(t) \end{bmatrix} \tag{4.96}$$

has been introduced. Sticking to classical literature on (discontinuous) Galerkin methods for ordinary differential equations (cf. e.g. [65], [35] and [36]), an approximate solution $\boldsymbol{z}^h$ to (4.94) is searched for in a finite dimensional subspace spanned by polynomial basis functions $P^k(\mathcal{T}_A)$ on subintervals $\mathcal{T}_A = [t_A, t_{A+1}] \subset \mathcal{T}$. Herein, P^k denotes the space of all polynomials of degree $\leq k$. For the sake of clarity, a uniform discretization $t_{A+1} = t_A + Ah$ for $0 \leq A \leq N$ and $Nh = T$ is assumed. Then the approximate solution $\boldsymbol{z}^h$ is searched for, such that

$$\int_{\mathcal{T}_A} \boldsymbol{w} \cdot \left(\mathrm{D}_t \boldsymbol{z}(t) + \boldsymbol{\mathcal{F}}(\boldsymbol{z}, t) + \boldsymbol{\mathcal{U}}\right) \; \mathrm{d}t + \boldsymbol{w}_A^+ \cdot [\![\boldsymbol{z}_A]\!] + \boldsymbol{w}_{A+1}^- \cdot \boldsymbol{\mathcal{H}} [\![\boldsymbol{z}_{A+1}]\!] = \boldsymbol{0} \tag{4.97}$$

is satisfied for all $\boldsymbol{w} \in P^k$. In Eq. (4.97) the abbreviations

$$\boldsymbol{z}_A^+ = \lim_{\xi \to 0} \boldsymbol{z}(t_A + \xi) \quad \textit{and} \quad \boldsymbol{z}_A^- = \lim_{\xi \to 0} \boldsymbol{z}(t_A - \xi) \tag{4.98}$$

has been introduced, denoting the limit of $\boldsymbol{z}$ from above and below, respectively. Furthermore, the bracket $[\![\boldsymbol{z}_A]\!] = \boldsymbol{z}(t_A^+) - \boldsymbol{z}(t_A^-)$ represents the jump of $\boldsymbol{z}$ at time t_A. An illustration thereof is depicted in Fig. 4.19 for $k = 1$.

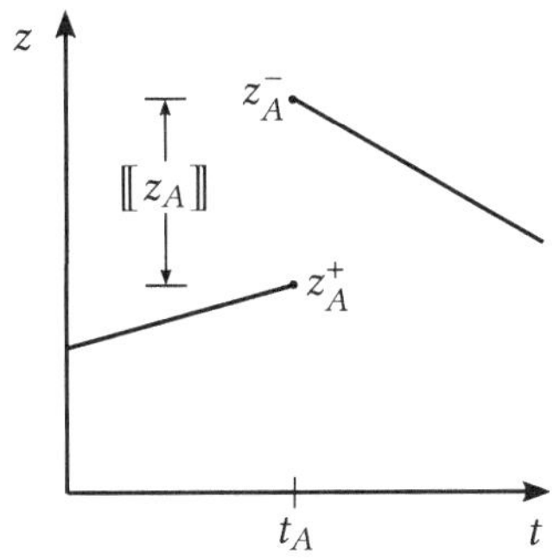

Figure 4.19.: Discontinuous approximation of the test and trial functions for $k = 1$.

The spaces of the test and trial functions $\boldsymbol{w}$ *and* $\boldsymbol{z}$ *are chosen to be the same. In general, test and trial functions are discontinuous at the nodes* t_A. *Consequently, initial conditions* $\boldsymbol{z}(t = 0) = \boldsymbol{z}_0$ *need to be imposed variationally by adding*

$$\boldsymbol{w}_A^- \cdot [\![\boldsymbol{z}_A]\!]$$

to Eq. (4.97). *A strong imposition would lead to an overdeterminacy of the coefficients of* $\boldsymbol{z}$. *Additionally, the servo-constraints* (4.95) *are imposed variationally*

$$\int_T \boldsymbol{w} \cdot (\boldsymbol{H}\boldsymbol{q} - \boldsymbol{\gamma}(t)) \, \mathrm{d}t = \boldsymbol{0} \tag{4.99}$$

by adding

$$\boldsymbol{w}_{A+1}^- \cdot \boldsymbol{\mathcal{H}} \cdot [\![\boldsymbol{z}(A+1)]\!] \tag{4.100}$$

to the left-hand side of (4.97), *where* $\boldsymbol{\mathcal{H}} = \begin{bmatrix} \boldsymbol{H}^T & \boldsymbol{0} \end{bmatrix}^T$ *has been introduced. This reflects the special structure of the inverse dynamics problem. Applying feasible interpolatory quadrature formulas, leads to a computational form of* (4.97) *and* (4.99).

4.3. Numerical investigations

This section aims to solve numerically the inverse dynamics problem of several mechanical systems, by applying the newly developed space-time integration methods. In particular, the method of characteristics, termed *MOC* (Sec. 4.1), and the space-time finite element method, termed *ST-FEM* (Sec. 4.2), are applied. ST-FEM relies on bi-linear isoparametric space-time elements. The evaluation of the integrals in (4.82) relies on standard Gaussian quadrature rules. This section partly reproduces [93].

Example 4.23 (Linear-elastic bar). *Starting with the feedforward control of the linear-elastic bar, enables a comparison of the numerical results delivered by both MOC and ST-*

FEM with the analytical solution (cf. Rem. 2.6 and Rem. 2.7). The prescribed trajectory of the bar at s = L is assumed as

$$
\gamma(t) = \begin{cases} 0 & t < t_0 \\ \dfrac{1}{2}\sin\left(\pi\dfrac{t-t_0}{t_f-t_0} - \dfrac{\pi}{2}\right) + \dfrac{1}{2} & t_0 \le t \le t_f \\ 1 & t > t_f \end{cases} \tag{4.101}
$$

for $t_0 = 1$ and $t_f = 3$. Moreover, the remaining data for this problem are $EA = 1$, $\rho A = 1$ and $L = 1$.

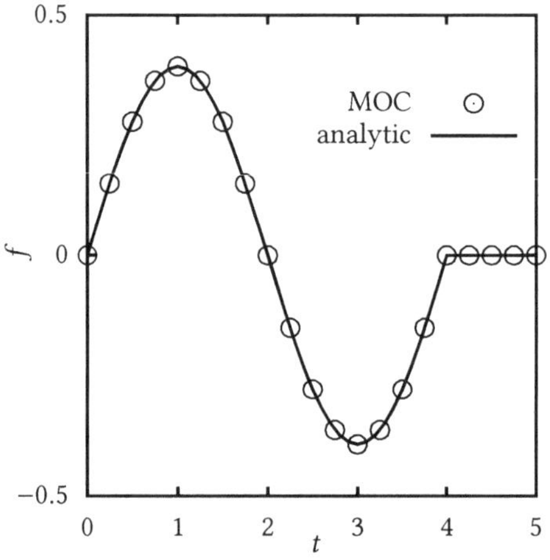

Figure 4.20.: Numerical solution for the actuating force $f(t)$ computed using the method of characteristics (circles) compared with the analytical solution (solid line).

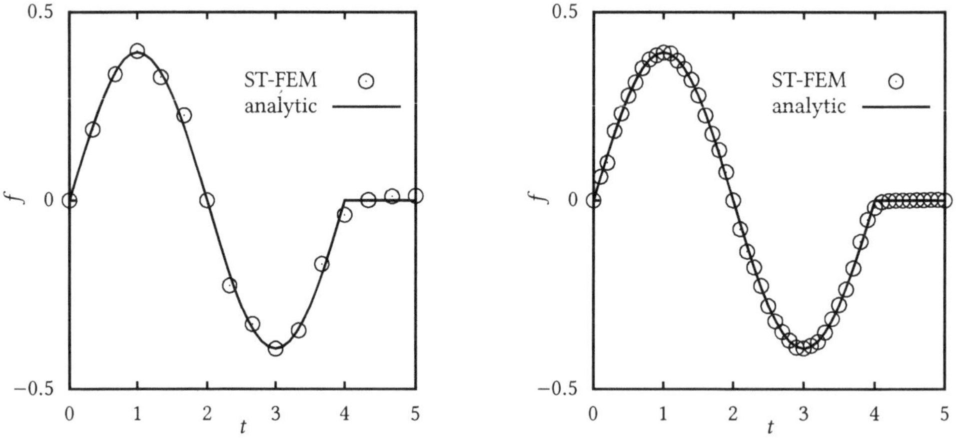

Figure 4.21.: Numerical solution for the actuating force $f(t)$ computed using the space-time finite element method (circles) compared with the analytical solution (solid line). Left diagram: $5 \times 15 = 75$ elements, right diagram: $10 \times 50 = 500$ elements.

In Fig. 4.20 and Fig. 4.21 the numerical results for the actuating force $f(t)$ are compared to the reference solution obtained by applying formula (2.22). It can be observed that the numerical results of the two alternative schemes under consideration closely match the reference solution.

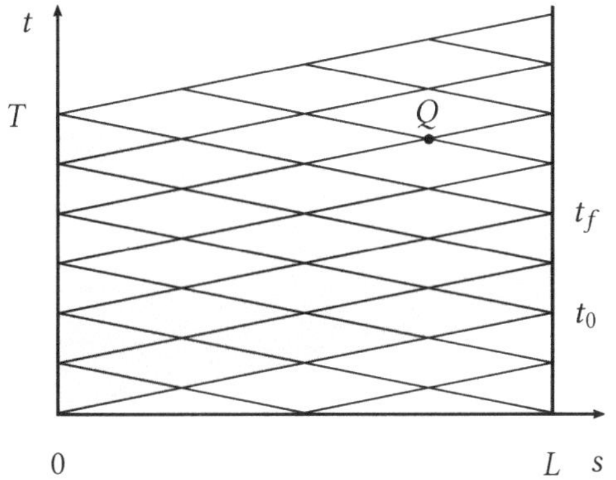

Figure 4.22.: Characteristic net along with a representative node Q located at $(s_Q, t_Q) \in \Omega$.

The results of MOC rely on a total of 356 unknowns. In Fig. 4.22, exemplarily a characteristc net is depicted. Note that due to the underlying hyperbolic structure, pre- and post-actuation phases have to be taken into account (cf. Rem. 4.9). Prescribing the motion on the boundary at $s = L$ during the time interval $[t_0, t_f]$ leads to the requirement of pre- and post-actuation phases in the solution of the inverse dynamics problem. This can be observed from Fig. 4.22 where the pre-actuation phase is related to $t \in [0, t_0]$ and the post-actuation phase is related to $t \in [t_f, T]$. Since the slope of the characteristic lines depends on the velocity of wave propagation, the span of both, pre- and post-actuation phases is determined by the length of the bar along with the the velocity of wave propagation. The shaded trapezoidal area in Fig. 4.22 elucidates this correlation.

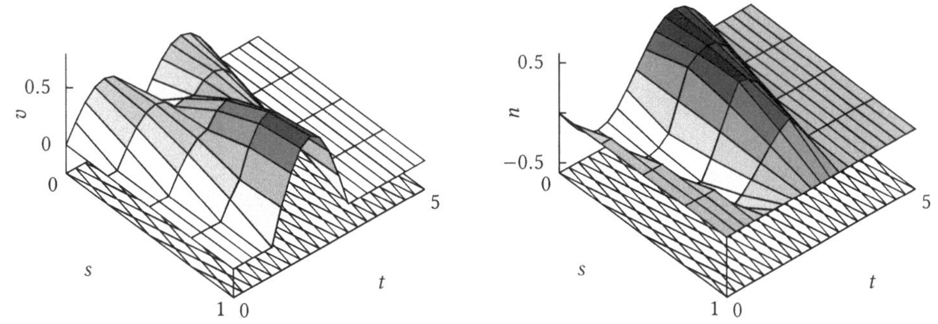

Figure 4.23.: Functions $v(s, t) = \partial_t u(s, t)$ and $n(s, t) = B\partial_s u(s, t)$ computed with the method of characteristics. Part of the characteristic net is also visualized in the s, t-plane.

Fig. 4.23 depicts the numerical solution for the two functions $v(s,t) = \partial_t r(s,t)$ *and* $n(s,t) = B\partial_s r(s,t)$*, respectively, computed with the method of characteristics. The corresponding characteristic net is also partially visualized in the* s,t*-plane.*

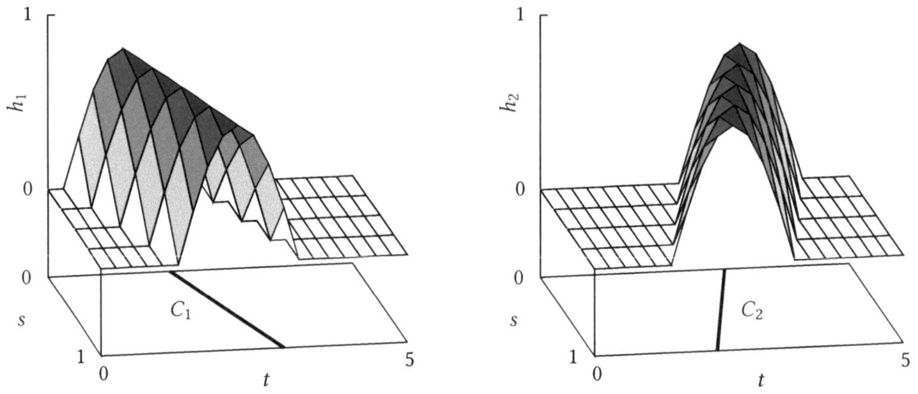

Figure 4.24.: Functions $h_\alpha(s,t)$ associated with the two Riemann Invariants.

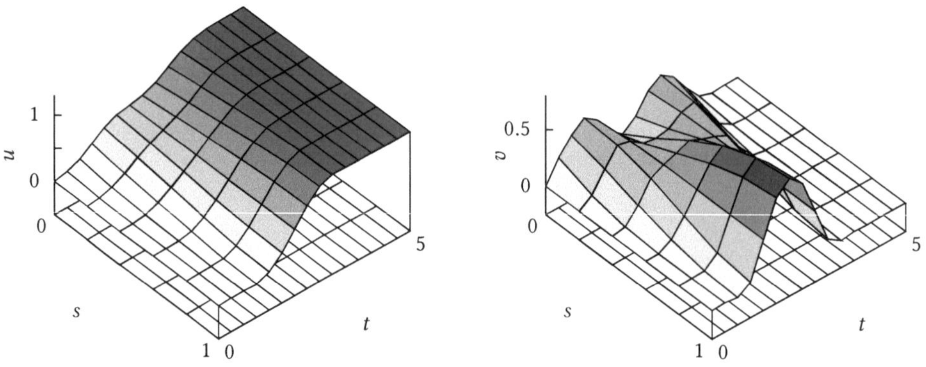

Figure 4.25.: Numerical solution for the displacement field $u(s,t)$ and velocity field $v(s,t)$ computed with the space-time finite element method. The corresponding finite element mesh (5×15 bilinear elements) can also be partially seen in the s,t-plane.

Concerning ST-FEM, two alternative meshes comprised of rectangular bi-linear elements have been applied. The first one consists of $5 \times 15 = 75$ *elements, leading to a total number of 195 unknowns. The second one consists of* $10 \times 50 = 500$ *elements, leading to a total* *number of 1150 unknowns. Fig. 4.24 contains plots of the two functions $h_\alpha(s,t)$ introduced in (4.47) associated with the Riemann Invariants (cf. Rem. 4.10). In particular, function $h_1(s,t)$ corresponds to the characteristic line C_1 whose slope is given by $ds/dt = k'(t) = c$. Similarly, function $h_2(s,t)$ corresponds to the characteristic line C_2 whose slope is given by $ds/dt = k'(t) = -c$. It can be observed that both functions are indeed constant along the respective characteristic line. The numerical solution for the displacement field $u(s,t)$ and the velocity field $v(s,t)$ computed with the space-time finite element method are depicted in Fig. 4.25. There, the corresponding finite element mesh is also partially visible in the s,t-plane.*

Concerning numerical accuracy, the method of characteristics yields exact results in the linear setting. This is due to the fact that the difference scheme presented in Rem. 4.8 is capable to exactly integrate along straight characteristic lines. In contrast to this, the accuracy of ST-FEM is of order two. This can be concluded from Fig. 4.26, in which the relative error

$$\varepsilon_f(h_e) = \frac{|f^{ana} - f^{h_e}|}{|f^{ana}|}$$

is shown. Here, f^{ana} is the actuating force calculated by applying the analytical solution (2.22), while f^{h_e} refers to the numerical solution obtained with ST-FEM. To get the results in Fig. 4.26, the space-time finite element mesh has been uniformly refined. In this connection, h_e denotes the element length in space direction.

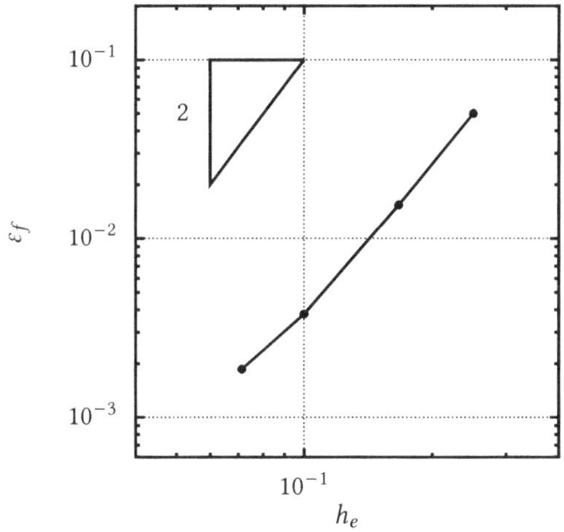

Figure 4.26.: Relative error in the actuating force f by using the space-time finite element method.

Example 4.24 (Geometrically exact string - planar formulation). *The next example deals with the planar motion of a geometrically exact string. In particular, the trajectory of the string at $s = L$ is prescribed as straight line in the x,y-plane via*

$$\boldsymbol{\gamma}(t) = \gamma(t)\begin{bmatrix}1\\1\end{bmatrix} \tag{4.102}$$

where function $\gamma(t)$ is given by (4.101) *with* $t_0 = 2$ *and* $t_f = 4$. *The remaining data is given by* $EA = 1$, $\rho A = 1$, $L = 1$ *and* $g = 9.81$.

The initial values for the inverse dynamics problem (2.11) *have been obtained by solving the equilibrium problem of the string by means of the semi-discrete model described in Sec. 3.1. For that purpose, 50 finite elements have been employed. The thus obtained vertical equilibrium configuration of the string is characterized by a total length* (2.6) *of* $l(0) = 9.985$ *and a suspension force of* $\boldsymbol{f}_0 = -g\,\boldsymbol{e}_y$.

Fig. 4.28 depicts the components of the actuating force required to realize the partially prescribed motion of the string. It can be observed that both the method of characteristics and the space-time finite element method yield closely related results. These results have been obtained by employing $50 \times 150 = 7500$ *rectangular space-time elements leading to a total number of* 15600 *unknowns. Merely 4 Newton iterations were necessary to achieve the present results by applying the simultaneous solution procedure outlined in Sec. 4.2.*

On the other hand, the results of the method of characteristics relies on a total number of 6727 *unknowns. Convergence was attained after 8 Newton iterations. It is worth noting that the data pertaining to the initial equilibrium configuration of the string has been used to initialize Newton's method throughout the space-time domain, both for ST-FEM and MOC.*

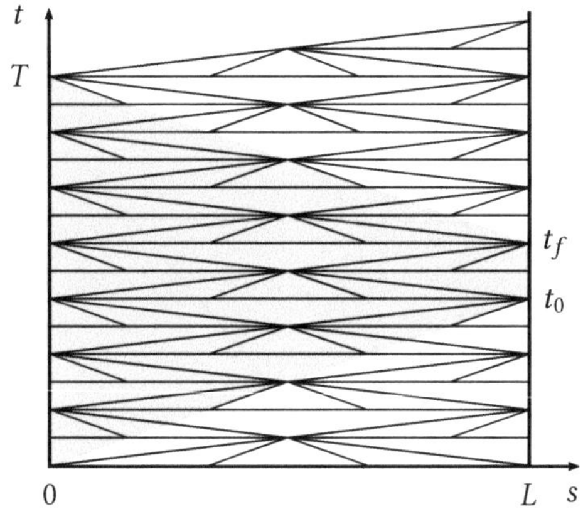

Figure 4.27.: Characteristic net for the geometrically exact formulation of strings.

In Fig. 4.27, exemplarily a charcateristc net is depicted. Note that similar to the linear setting, the solution of the inverse dynamics problem requires pre- and post-actuation phases. This is again indicated with the shaded area in Fig. 4.27. The parameters t_0, t_f *and* T *have the same meaning as in Ex. 4.23.*

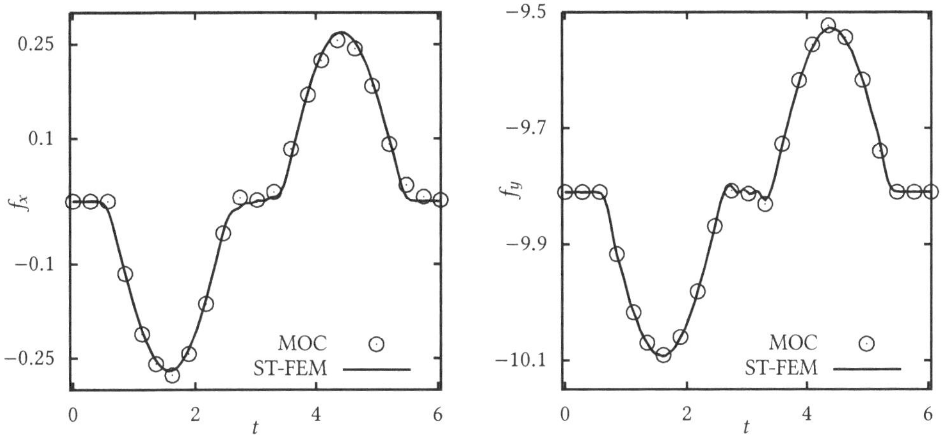

Figure 4.28.: Actuating force components $f_x(t)$ and $f_y(t)$ at $s = 0$ accounting for a partly prescribed motion of the string at $s = L$.

To get an impression of the overall motion of the string, snapshots in the x, y-plane and in space-time are shown in Fig. 4.29 and Fig. 4.30, respectively. In addition to that, the stretch distribution over the space-time domain is depicted in Fig. 4.31.

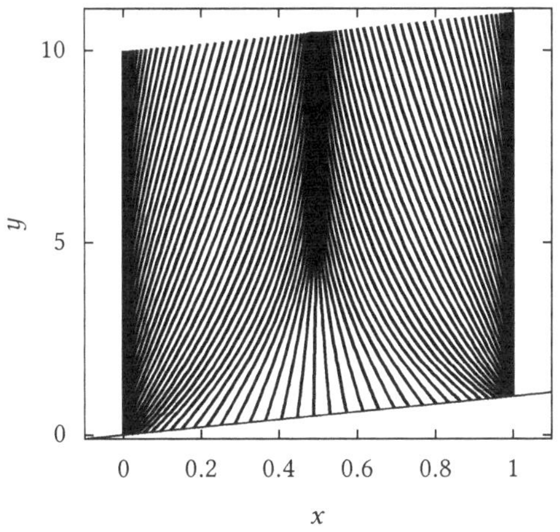

Figure 4.29.: Snapshots of the planar motion of the string in the x, y-plane.

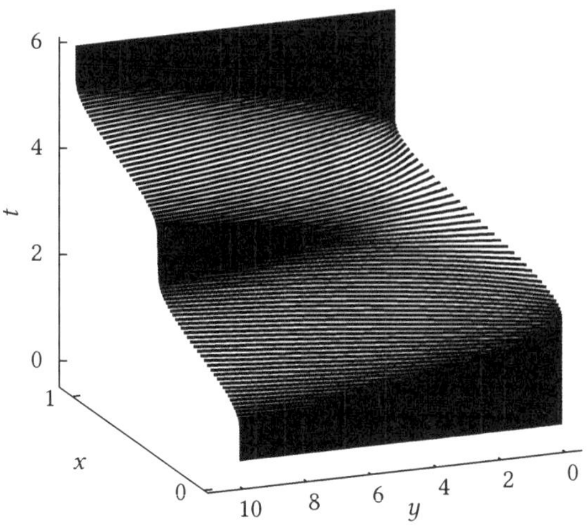

Figure 4.30.: Snapshots of the motion of the string in the space-time domain.

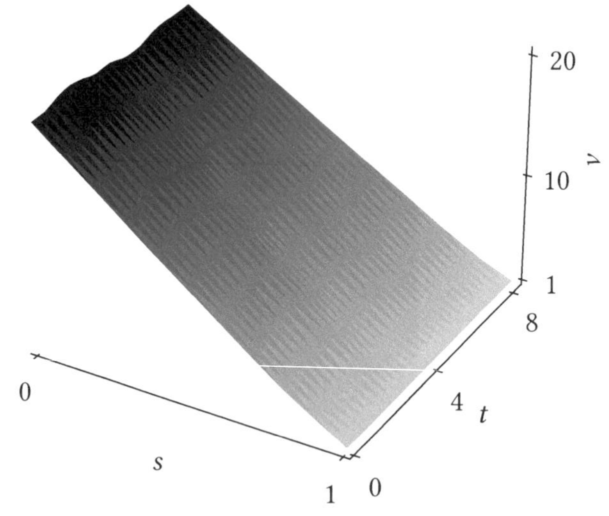

Figure 4.31.: Stretch distribution over the space-time domain.

Example 4.25 (Additional mass). *In extension to Ex. 4.24, an additional point mass* $M = 1$ *is attached to the right end of the string (cf. Rem. 2.2 as well as Rem. 3.2), then the term* $\boldsymbol{\eta}(t) = M\left(\boldsymbol{g} - \partial_t \boldsymbol{v}^h(L,t)\right)$ *has to be taken into account in weak form* (4.80)$_2$. *Correspondingly, in the discrete formulation, the following entries of matrices* (4.82)$_1$ *and* (4.82)$_5$ *need be modified according to*

$$\mathcal{A}_{ab} \leftarrow \mathcal{A}_{ab} + \boldsymbol{I}_d\, M \int_{\partial\Omega_\eta} N_a \partial_t N_b \,\mathrm{d}t$$

$$\mathcal{F}_a^{\mathrm{ext}} \leftarrow \mathcal{F}_a^{\mathrm{ext}} + M\boldsymbol{g} \int_{\partial\Omega_\eta} N_a \,\mathrm{d}t$$

for $a, b \in \mathcal{N}_\eta$. *The prescribed motion of the point mass is again governed by* (4.102), *with* $t_0 = 1$ *and* $t_f = 3$. *The remaining data is specified by* $EA = 10$, $\rho A = 1$, $L = 1$ *and* $g = 9.81$.

In a first step the equilibrium configuration of the string-mass system has been computed by applying 50 finite elements within the semi-discrete formulation (Sec. 3.1). Accordingly, the vertical equilibrium configuration providing the initial data for the subsequent inverse dynamics simulation is characterized by a total length (2.6) *of* $l(0) = 3.258$ *and a suspension force of* $\boldsymbol{f}_0 = -2g\,\boldsymbol{e}_y$.

The resulting components of the actuating force required to realize the partially specified motion are depicted in Fig. 4.32.

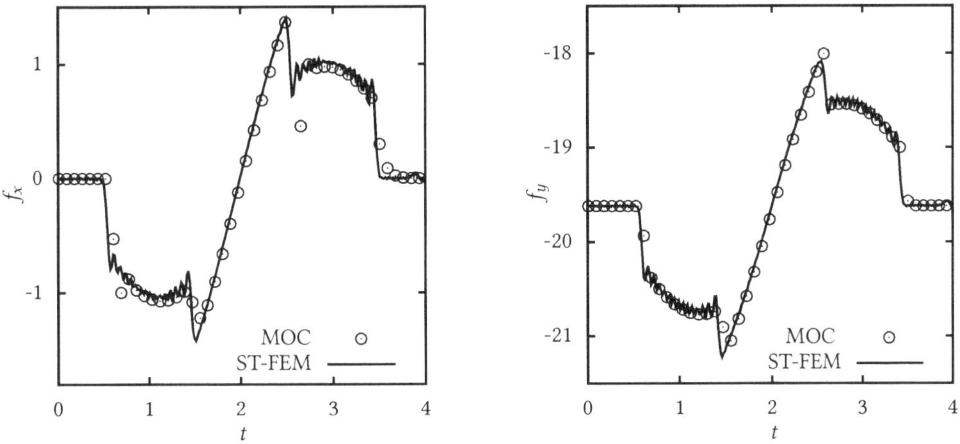

Figure 4.32.: Actuating force components $f_x(t)$ and $f_y(t)$ at $s = 0$ accounting for a prescribed motion of a mass point attached to the string at $s = L$.

It can be seen that the results of both methods under investigation match each other closely. The numerical results of the space-time finite element method are based on a mesh containing $50 \times 200 = 10000$ *elements leading to a total number of 20800 unknowns. Merely 4*

Newton iterations are required to attain the converged solution. The corresponding results of the method of characteristics are based on a total number of 10515 unknowns. The required number of Newton iterations amounts to 7. Again the data pertaining to the initial equilibrium configuration of the system has been used to initialize Newton's method throughout space-time.

The overall motion of the string is illustrated with snapshots of successive configurations of the system in Fig. 4.33. The total length (2.6) of the string versus time is depicted in Fig. 4.34. In addition to that, the stretch distribution over the space-time domain is depicted in Fig. 4.35.

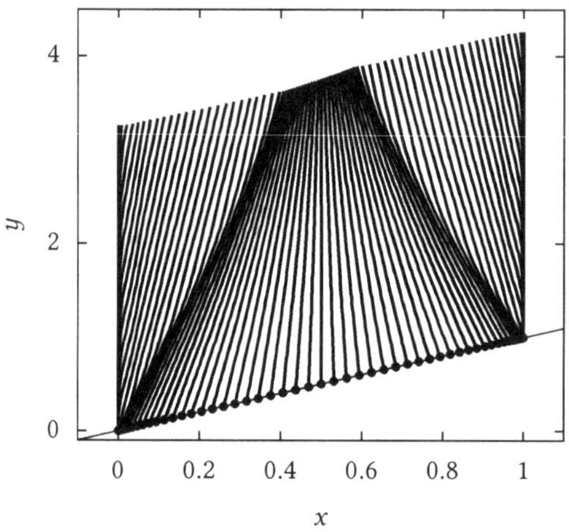

Figure 4.33.: Snapshots of the motion of the string with attached mass point.

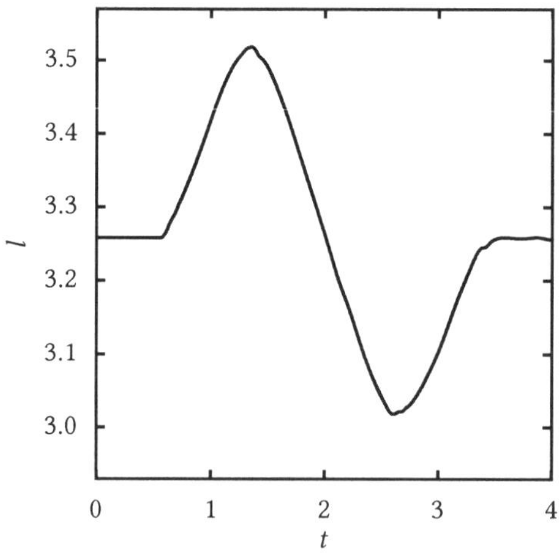

Figure 4.34.: Total length of the string.

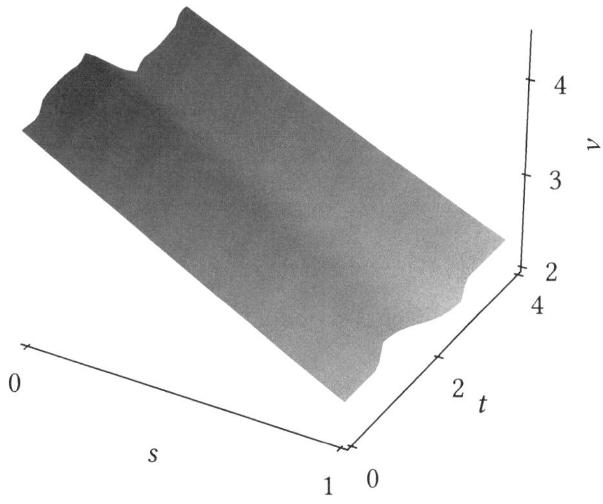

Figure 4.35.: Stretch distribution of a string with attached mass point over the space-time domain.

Example 4.26 (Geometrically exact string). *This example aims to underline the importance of the proposed method for the feedforward control of the three-dimensional motion of the string. In particular, a point mass, attached at $s = L$ (cf. Rem. 2.2, Rem. 3.2 and Ex. 4.25) is supposed to move on a helix. To this end, the helix-shaped trajectory is prescribed by*

$$\boldsymbol{y}(t) = \begin{pmatrix} \cos(2\pi \cdot \varphi) - 1 \\ \sin(2\pi \cdot \varphi) \\ z_f \varphi \end{pmatrix} \forall t \in I, \quad \boldsymbol{y}(t) = \begin{pmatrix} 0 \\ 0 \\ 0 \end{pmatrix} \forall t < t_0, \quad \boldsymbol{y}(t) = \begin{pmatrix} 0 \\ 0 \\ z_f \end{pmatrix} \forall t > t_f \tag{4.103}$$

where

$$\varphi = -\frac{1}{2}\cos(\pi \cdot y) + \frac{1}{2} \qquad y = \frac{1}{2}\sin\left(\pi \frac{t - t_0}{t_f - t_0} - \frac{\pi}{2}\right) + \frac{1}{2} \qquad I = \left[t_0, t_f\right] = [2, 8]$$

In Fig. 4.36 the prescribed coordinates are plotted versus time. The remaining data for this example is specified by $EA = 10$, $\rho A = 1$, $L = 1$, $g = 9.81$, $M = 1$ and $z_f = 5$.

In a first step the vertical equilibrium configuration of the system (Fig. 4.37) subject to gravity load is computed by applying the semi-discrete model of the string (cf. Sec. 3.1) consisting of 15 finite elements. Accordingly, the vertical equilibrium configuration is characterized by a total length (2.6) of $l(0) = 3.258$ and a suspension force of $\boldsymbol{f}_0 = -2g\boldsymbol{e}_z$. The thus obtained equilibrium configuration provides the initial data for the inverse dynamics problem (2.11) to be solved subsequently. In this connection, the data is also used to initialize Newton's method through the space-time domain. The space-time finite element approach (Sec. 4.2) is applied in conjunction with the simultaneous solution procedure outlined in Sec. 4.2.

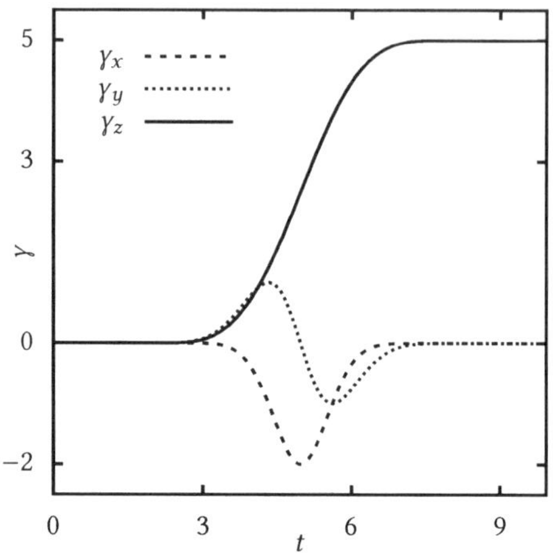

Figure 4.36.: Components of the prescribed helix.

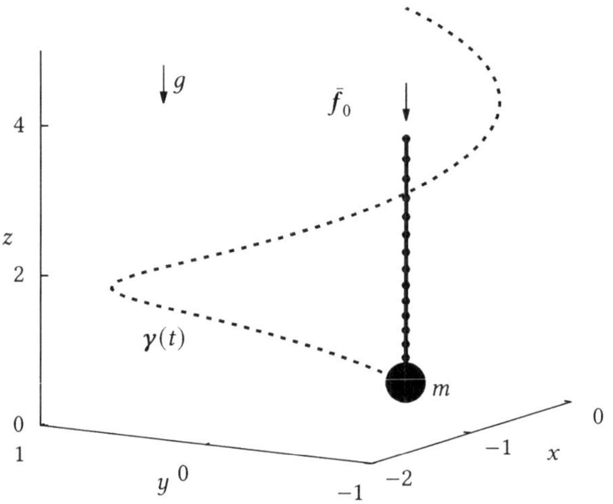

Figure 4.37.: Equilibrium configuration of the string with attached point mass under gravity load. In addition to that, the helix-shaped trajectory to be tracked by the point mass is indicated as well.

The computed three components of the actuating force required to realize the partially prescribed motion of the system are depicted in Fig. 4.38. Furthermore, Fig. 4.39 displays the total length of the string versus time. The resulting motion of the system is illustrated with Fig. 4.40 and Fig. 4.41. These results have been obtained by employing $15 \times 149 = 2235$ *rectangular space-time finite elements amounting to a total number of 7599 unknowns. Merely 4 Newton iterations were required to reach the numerical solution.*

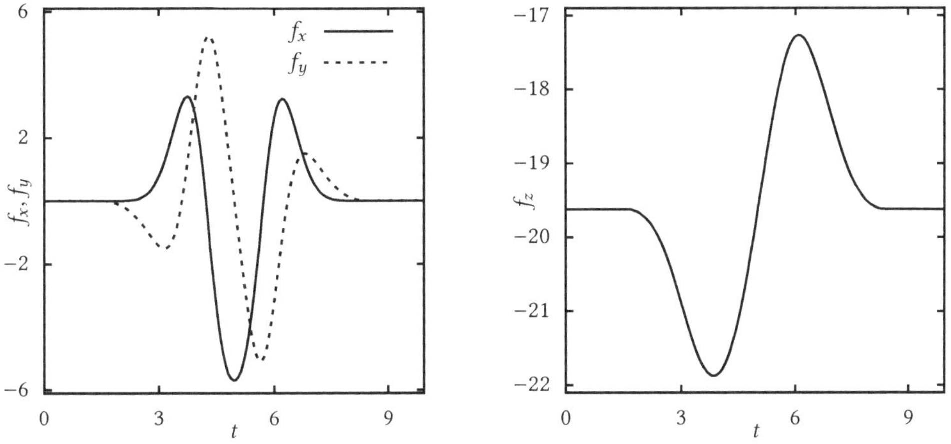

Figure 4.38.: Components of the actuating force $\boldsymbol{f}(t)$ obtained as solution of the inverse dynamics problem.

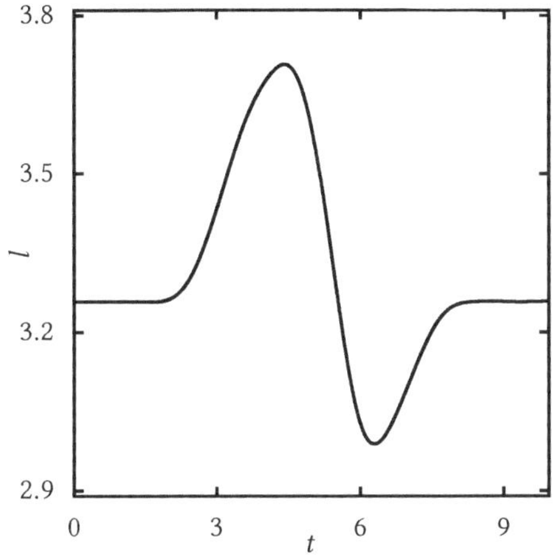

Figure 4.39.: Total length of the string.

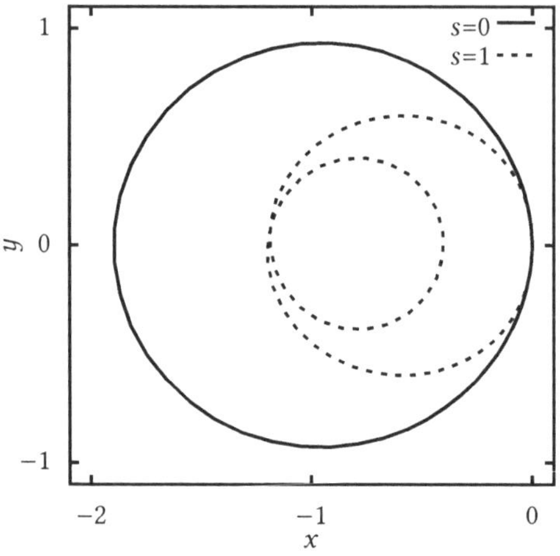

Figure 4.40.: Computed trajectory of the actuated upper end of the string ($s = 0$) projected onto the horizontal x, y-plane. In addition to that, the projection of the prescribed trajectory of the lower end of the string ($s = 1$) is also shown.

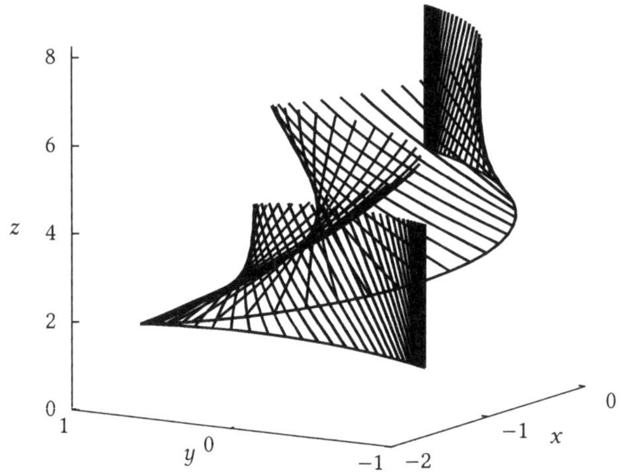

Figure 4.41.: Snapshots of the moving elastic string (the point mass attached at the lower end of the string is not shown). According to the inverse dynamics problem solved, the lower end of the string starts at rest (starting point $\boldsymbol{\gamma}(t_0) = (0, 0, 0)$) and traces the prescribed helix-shaped trajectory until the end point $\boldsymbol{\gamma}(t_f) = (0, 0, 5)$ has been reached.

Example 4.27 (Recursive solution procedure). *We next apply the recursive solution procedure described in Sec. 4.2. The data is kept the same as before. The number of finite elements in space direction is reduced from 15 to 10 in order to simplify the graphical illustration of the results. Accordingly, the space-time finite element mesh is comprised of 10 × 149 rectangular elements leading to* $N = 10$ *time-space slabs used in the recursive solution procedure. As expected, the numerical results are practically indistinguishable from those obtained by applying the simultaneous solution procedure (cf. Sec. 4.2). In Fig. 4.42 the contact forces and* *trajectories along the boundaries Γ_n of the time-space slabs are shown. While the dimension of the iteration matrix pertaining to the simultaneous solution is equal to 5364, the dimension of the iteration matrix of the recursive solution procedure is equal to 1341 (cf. Rem. 4.20). On average 3.2 Newton iterations were required in each of the 10 steps of the recursive scheme. In comparison, 4 Newton iterations were required to reach the solution of the simultaneous solution scheme.*

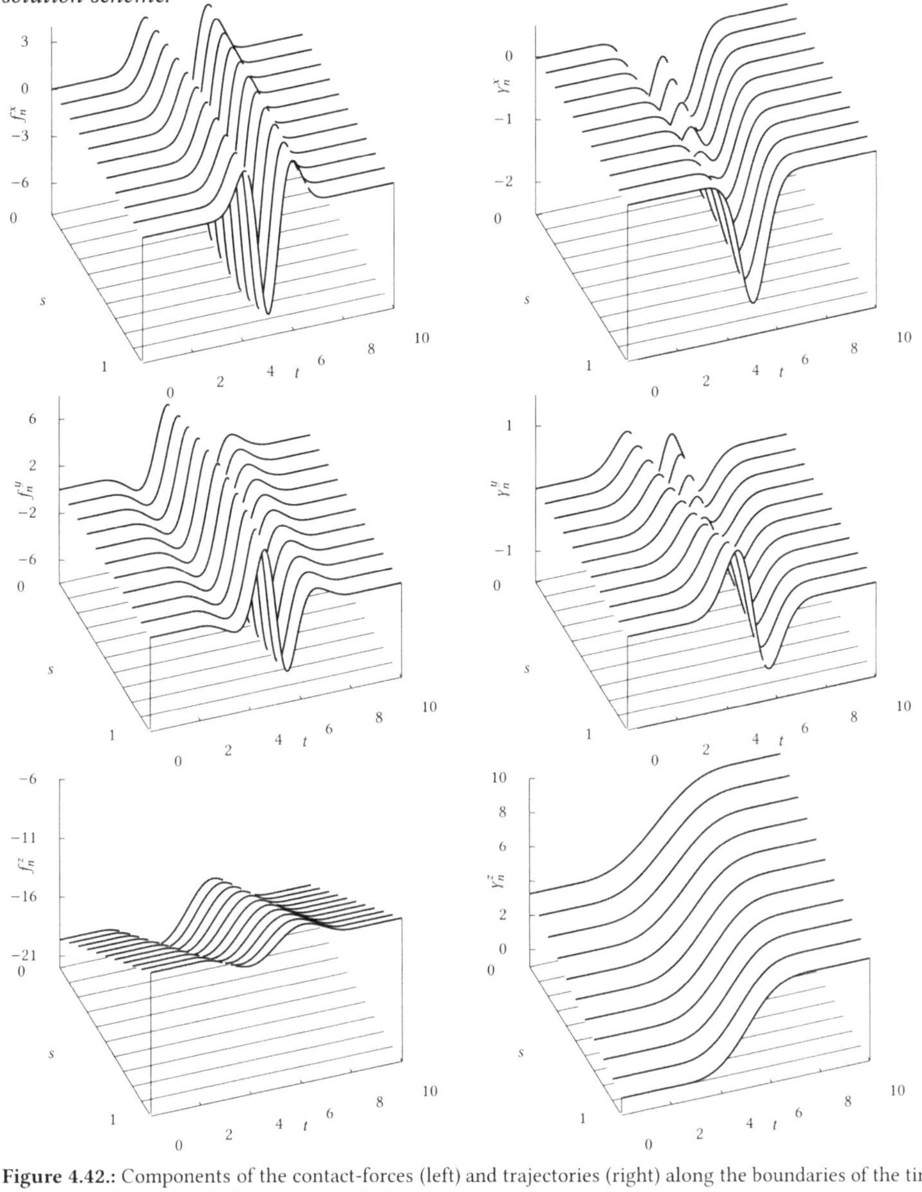

Figure 4.42.: Components of the contact-forces (left) and trajectories (right) along the boundaries of the time-space slabs obtained with the recursive solution procedure.

Remark 4.28 (Perturbed data of the prescribed trajectory). *According to Hadamard, a problem is well-posed if there exists a solution to the problem which is unique and stable (see [45] or [58, Def. 1.13]). Since existence and uniqueness of the solution to the present space-time boundary value problem are assured by the differential flatness of the system, stability still has to be ensured. The solution is called stable if it depends continuously on the given data - small perturbations of the prescribed servo-constraints on $\partial\Omega_\eta$ must not lead to unbounded results for the actuating force $\boldsymbol{f}(t)$ on $\partial\Omega_f$. This can be verified numerically by adding random noise to the given data of the trajectory such that*

$$\left\|\boldsymbol{\gamma}^\delta - \boldsymbol{\gamma}\right\|_2 \leq \delta$$

where $\boldsymbol{\gamma}^\delta$ denotes the perturbed trajectory. The perturbed components of $\boldsymbol{\gamma}^\delta$ are shown in Fig. 4.43. For comparison, the original components of $\boldsymbol{\gamma}$ given by (4.103) *are depicted in Fig. 4.36.*

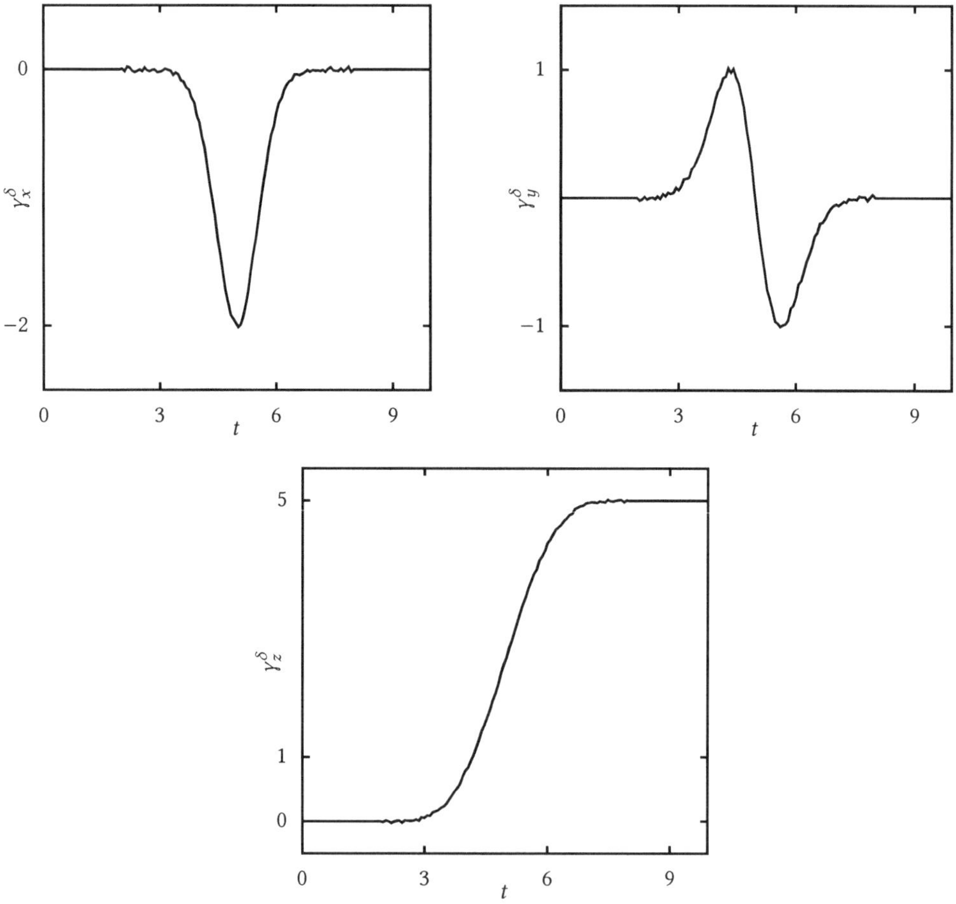

Figure 4.43.: Components of the perturbed data for the prescribed trajectory $\boldsymbol{\gamma}^\delta$ for $\delta : t \mapsto [0, 0.05]$.

Fig. 4.44 shows the results for the actuating forces resulting from the solution of the inverse problem with the perturbed data. Comparison with Fig. 4.38 indicates that the perturbed problem still yields results close to the original ones.

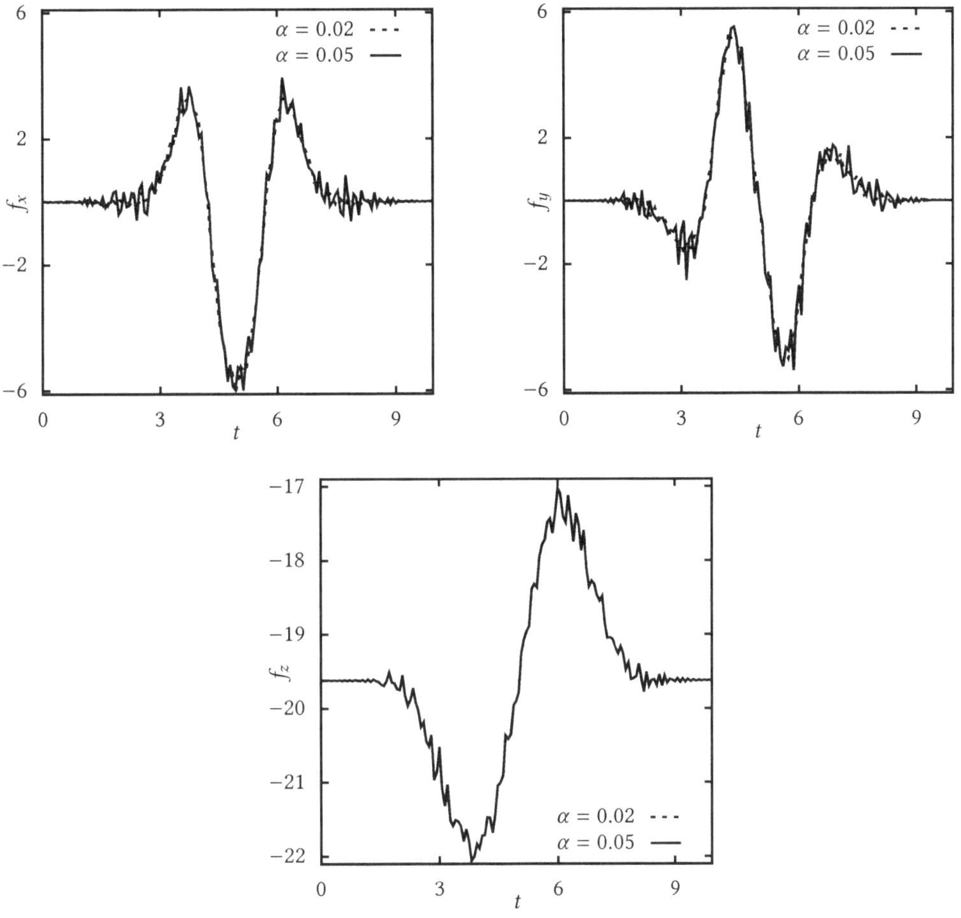

Figure 4.44.: Components of the actuating force obtained as solution of the inverse problem with noisy data for the prescribed trajectory $\boldsymbol{\gamma}^{\delta}$ at $\partial\Omega_{\eta}$ with $\delta : t \mapsto [0, \alpha]$ and $\alpha \in \{0.02, 0.05\}$.

Example 4.29 (Geometrically exact beam formulation - circular trajectory). *A beam, with mass density $\rho = 1$ and axial-, bending- and shear stiffness $EA = 1, EI = 1$ and $GA = 1$ respectively, is investigated. The length of the beam in a stress-free reference configuration is assumed to be $L = 1$. The aim is, to find the actuation $\boldsymbol{f} = \begin{bmatrix} f_x & f_y & m \end{bmatrix}^T$ applied at $s = 0$,* *such that the end-effector of the beam at* $s = L$ *follows a prescribed trajectory. For this, a circular motion starting at* $t_0 = 2$ *and ending at* $t_f = t_0 + T = 4$ *is choosen as given by*

$$\boldsymbol{\gamma} = \begin{bmatrix} 1 \\ 0 \\ 0 \end{bmatrix} \forall t < t_0 \qquad \boldsymbol{\gamma} = \begin{bmatrix} 1 - \cos(\varphi \cdot y(t)) \\ \sin(\varphi \cdot y(t)) \\ \varphi \cdot y(t) \end{bmatrix} \forall t \in [t_0, t_0 + T] \qquad \boldsymbol{\gamma} = \begin{bmatrix} 0 \\ 1 \\ \varphi \end{bmatrix} \forall t > t_f. \tag{4.104}$$

Herein $\varphi \in [0, 2\pi]$ *has been introduced, specifying the angle of rotation. Furthermore, the function* $y : \mathcal{T} \mapsto \mathbb{R}$ *is obtained as*

$$y(t) = 1 - \frac{1}{2}\left(\cos\left(\frac{\pi}{2}\left(\sin\left(\pi\frac{t - t_0}{T} - \frac{\pi}{2}\right) + 1\right)\right) + 1\right).$$

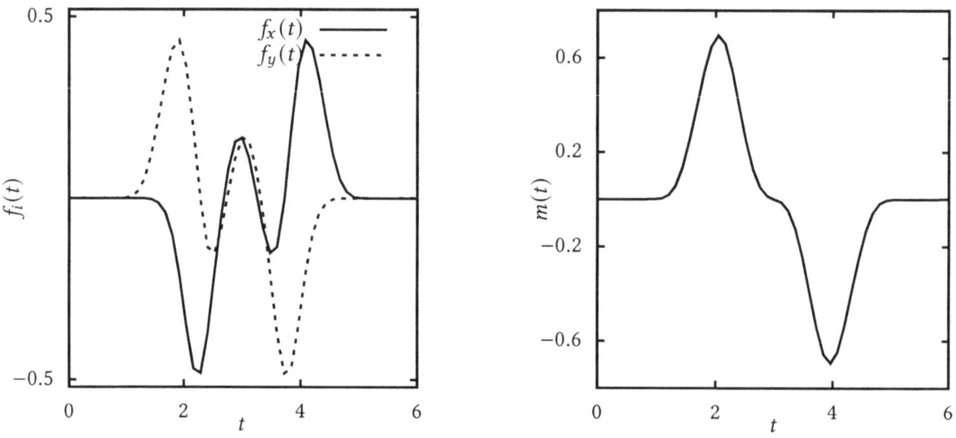

Figure 4.45.: Components of the force (left) and torque (right) acting at $s = 0$ computed with the proposed space-time finite element method such that the beam at $s = L$ follows a prescribed circle from $P_0(1, 0)$ to $P_T(0, 1)$.

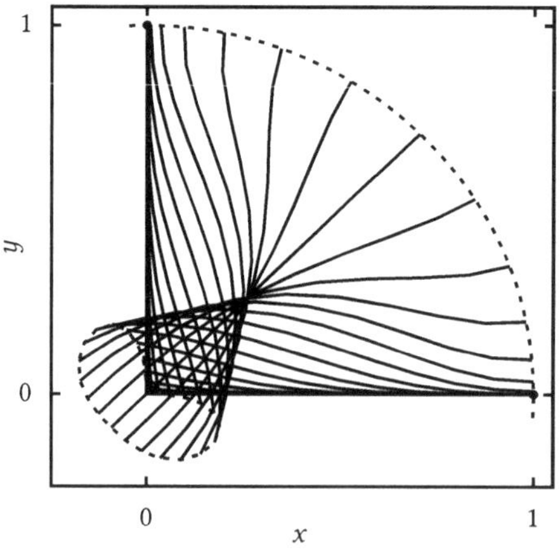

Figure 4.46.: Snapshots of the motion of a geometrically exact beam actuated at $s = 0$ such that its end at $s = 1$ aligns with a prescribed circular motion.

In Figure 4.45 the components of the actuating force f_i (left) and the actuating torque m (right) is depicted. Note that also the delay time can be observed herein. This is due to the hyperbolic structure of the underlying system mentioned earlier. In Figure 4.46 snapshots of the planar motion of the beam satisfying the prescribed trajectory at $s = L$ are shown. These results have been obtained by employing 10×50 *rectangular space-time elements leading to a total number of* 3450 *unknowns.*

Example 4.30 (Geometrically exact beam formulation - straight trajectory). *Considering a beam with material and geometrical properties taken from Ex. 4.29, is forced at $s = 0$ such that the beam at s=1 follows the prescribed motion*

$$\gamma = \begin{bmatrix} 1 \\ 0 \\ 0 \end{bmatrix} \forall t < t_0 \qquad \gamma = \begin{bmatrix} 1 + y(t) \\ y(t) \\ 0 \end{bmatrix} \forall t \in [t_0, t_0 + T] \qquad \gamma = \begin{bmatrix} 2 \\ 1 \\ 0 \end{bmatrix} \forall t > t_f. \tag{4.105}$$

where the function $y : \mathcal{T} \mapsto \mathbb{R}$ is assumed as

$$y(t) = 1 - \frac{1}{2}\left(\cos\left(\frac{\pi}{2}\left(\sin\left(\pi\frac{t - t_0}{T} - \frac{\pi}{2}\right) + 1\right)\right) + 1\right).$$

A simultaneous space-time integration of the partial differential equation at hand leads to the actuating force and torque at $s = 0$, searched for.

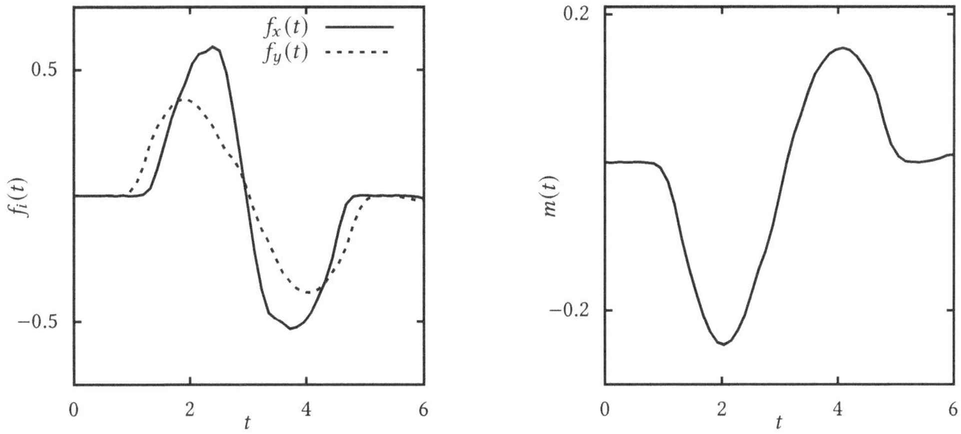

Figure 4.47.: Components of the force (left) and moment (right) acting at $s = 0$ computed with the proposed space-time finite element method such that the beam at $s = L$ follows a prescribed straight line from $P_0(1, 0)$ to $P_T(2, 1)$.

In Fig. 4.47, the force components f_x and f_y as well as the torque m are depicted. These results have been obtained by employing 10×50 *rectangular space-time elements leading to a total number of* 3450 *unknowns.*

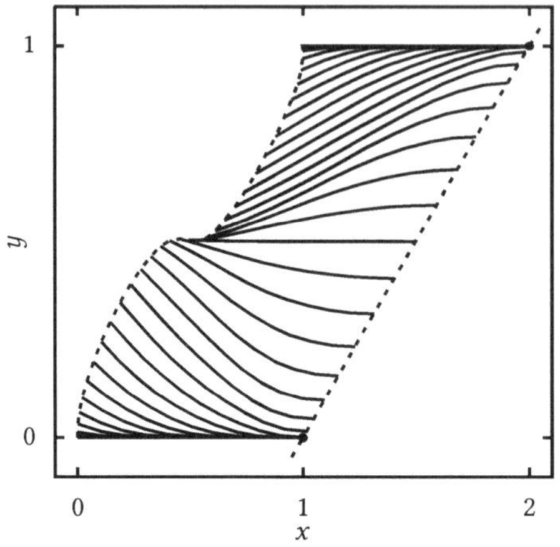

Figure 4.48.: Snapshots of the motion of a geometrically exact beam actuated at $s = 0$ such that its end at $s = 1$ aligns with a prescribed straight motion.

Example 4.31 (Nonlinear continuum - plane stress). *This example aims to demonstrate the relevance of the newly proposed simultaneous space-time integration of the inverse dynamics problem. For this, a quadratic membrane having edge length $l = 0.5$, mass density $\rho = 1$, Young's Modulus $E = 100$ and Poissons ratio $v = 0.3$ is considered. The actuation $\boldsymbol{f} : \partial\Omega_f \times \mathcal{T} \mapsto \mathbb{R}^n$ is searched for, such that $\boldsymbol{\gamma} : \partial\Omega_\eta \times \mathcal{T} \mapsto \mathbb{R}^n$ can be prescribed as*

$$\boldsymbol{\gamma} = \begin{bmatrix} 1 \\ s_2 \\ 0 \end{bmatrix} \forall t < t_0 \qquad \boldsymbol{\gamma} = \begin{bmatrix} 1 + y(t) \\ y(t) \\ 0 \end{bmatrix} \forall t \in [t_0, t_0 + T] \qquad \boldsymbol{\gamma} = \begin{bmatrix} 2 \\ s_2 + 1 \\ 0 \end{bmatrix} \forall t > t_f \,. \tag{4.106}$$

In Eq. (5.13), *the function $y : \mathcal{T} \mapsto \mathbb{R}$ is given by*

$$y(t) = 1 - \frac{1}{2}\left(\cos\left(\frac{\pi}{2}\left(\sin\left(\pi\frac{t - t_0}{T} - \frac{\pi}{2}\right) + 1\right)\right) + 1\right)$$

Note that the fully boundary-actuated problem, that has been considered in Ex. 4.23, Ex. 4.24, Ex. 4.25, Ex. 4.26, Ex. 4.29 and Ex. 4.30, is now extended to partly boundary-actuated problems, e.g. $\partial S_f \cup \partial S_\eta \neq \partial S$ is explicitly permitted.

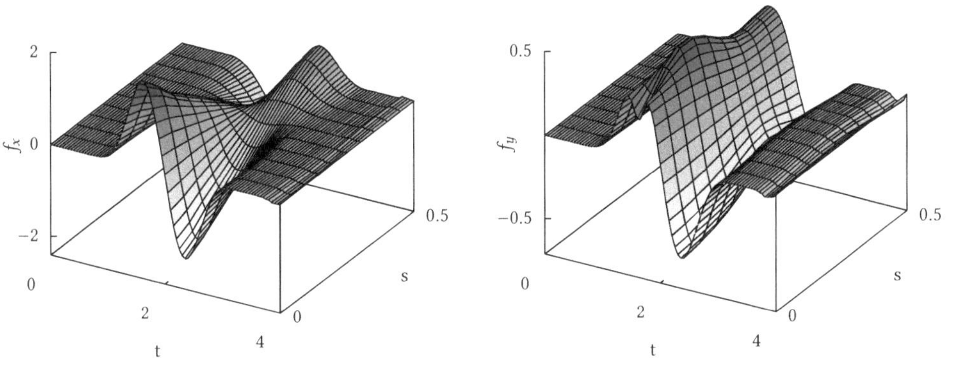

Figure 4.49.: Inverse computation of the actuating traction on $\partial\Omega_f$ accounting for a partly prescribed motion on $\partial\Omega_\eta$.

In Fig. 4.49 the components of the actuating force $f_i : \partial\Omega_f \times \mathcal{T} \mapsto \mathbb{R}$, computed using the space-time finite element method, are depicted. In Fig. 4.50 snapshots of the prescribed maneuver are illustrated. These results have been obtained by employing $10\times10\times80$ *hexahedral space-time elements leading to a total number of* 40480 *unknowns.*

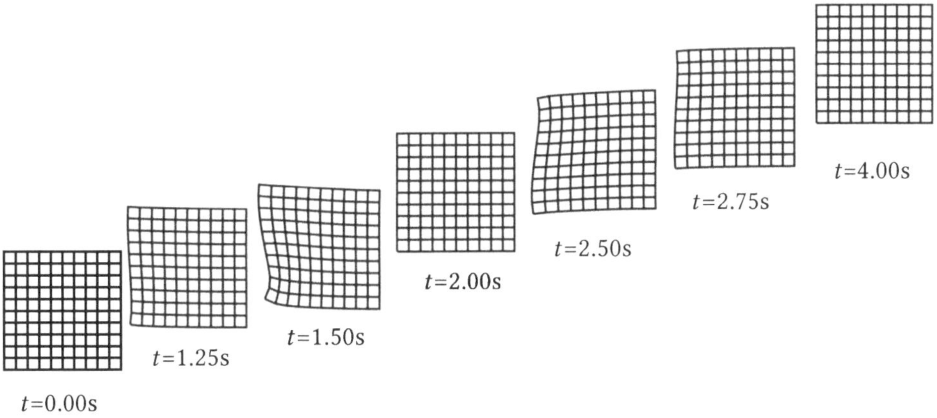

Figure 4.50.: Snapshots of the membrane actuated on $\partial\Omega_f$ such that the boundary $\partial\Omega_\eta$ translates by $\boldsymbol{u} = \begin{bmatrix}1 & 1 & 0\end{bmatrix}^T$ by simultaneously keeping its shape straight.

5. Flexible multibody systems

After the inverse dynamics problem for flexible mechanical systems, including slender structures such as strings and beams but also general three dimensional continua undergoing large deformations, has been analyzed, simultaneous space-time integration techniques could have been established in order to solve the control problem at hand. In the present Chapter, we aim to apply the newly established theory to the feed-forward control of flexible multibody systems. Therefore, the newly proposed simultaneous space-time integration is first adopted in Sec. 5.1 to the cooperative control problem of rigid bodies actuated through several flexible strings. Later on in Sec. 5.2, a rotational arm model consisting of *two* rigid and *one* flexible arm is investigated. Accompanying numerical examples are underlining the relevance of the proposed methods in context of flexible multibody systems.

5.1. Cooperative control problem

This section is concerned with finding the feed-forward control of rigid bodies actuated through multiple elastic strings. More precisely, *four* elastics strings are attached to a rigid body. The task is to find the actuating forces that are applied to the free end of each rope such that the rigid body follows a prescribed motion. To avoid singularities, we presume that the contact points do not lie on a plane, a straight line or are concentrated in one point. Instead we assume the contact points to be vertices of a tetrahedron. See Fig. 5.1 for an illustration of the control task.

The structure of the equations of motion of the underlying multibody system, comprised of geometrically exact ropes and rigid bodies, is outlined first. Numerical examples will be given underlining the applicability of the proposed solution strategy.

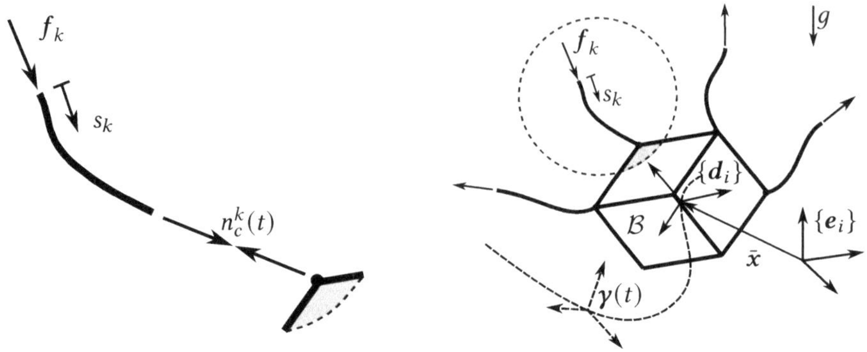

Figure 5.1.: Elastic string (left) as sub-system of a multibody system (right): Cooperative transportation of a rigid body through four elastic strings actuated by external forces $\boldsymbol{f}_k$, $k \in \{1, \ldots, 4\}$. In the inverse dynamics problem, the motion of the rigid body is prescribed and the goal is to determine the motion of the whole multibody system along with the actuating forces $\boldsymbol{f}_k$. The prescribed motion of the rigid body gives rise to the boundary conditions (trajectory and contact force $\boldsymbol{n}_c^k$) for each string at its point of attachment to the rigid body.

Rigid body. When considering general rigid bodies, in principle the same strategy as for the attached mass point can be applied (cf. Rem. 2.2 and Ex. 4.24 as well as Rem. 2.3). The actuating forces needed to achieve the desired motion of the rigid body can then be calculated directly by considering the governing equations of motion. For that purpose, the rigid body is modelled as a *Cosserat* continuum subjected to geometric constraints. Consequently, the motion of the rigid body is fully described by the directors $\boldsymbol{d}_i : T \mapsto \mathbb{R}^3$ and the position of the centre of gravity $\bar{\boldsymbol{x}} : T \mapsto \mathbb{R}^3$. Furthermore, the rigid body is assumed to have total mass $M = \int_{\mathcal{B}_0} \rho_0 \, \mathrm{dV}$ with a density function $\rho_0 : \mathcal{B}_0 \mapsto \mathbb{R}$. Accordingly, the equations of motion can be written as

$$
\begin{aligned}
\bar{M}\partial_t^2 \bar{\boldsymbol{x}} - \boldsymbol{f}_{ext} - \boldsymbol{G} &= 0 \, , \\
E_{ij}\partial_t^2 \boldsymbol{d}_j - \boldsymbol{f}_{ext}^i + \Lambda_{ij}\boldsymbol{d}_j &= 0 \, , \\
g_c(\boldsymbol{d}_i) = \boldsymbol{d}_i \cdot \boldsymbol{d}_j - \delta_{ij} &= 0 \, .
\end{aligned}
\tag{5.1}
$$

Herein $\boldsymbol{G} = \bar{M} g \boldsymbol{e}_z$ is the gravitational force and E_{ij} are the components of the referential Euler tensor which is closely related to the classical inertia tensor of rigid body dynamics (see [21] for more details). Furthermore, $\boldsymbol{f}_{ext}$ is the resultant external force and

$$
\boldsymbol{f}_{ext}^i = X_i^k \boldsymbol{n}_c^k \tag{5.2}
$$

are the external director forces (Fig. 5.2). This holds for a discrete actuation of the rigid body. Note that by introducing the matrix

$$
\mathcal{H}_{jk} = \begin{bmatrix} \boldsymbol{I} \\ X_1^k \boldsymbol{I} \\ X_2^k \boldsymbol{I} \\ X_3^k \boldsymbol{I} \end{bmatrix} , \tag{5.3}
$$

containing the information of the geometric position of the contact points, the following linear relation between the external (director) forces $\bar{f}(t) = \begin{bmatrix} f_{ext} & f^i_{ext} \end{bmatrix}^T$ and the contact forces $n^k_c(t)$ is obtained as

$$\bar{f}(t) = \mathcal{H} n^k_c(t) \,. \tag{5.4}$$

The geometric constraints (5.1)$_3$ are enforced by the Lagrange multipliers Λ_{ij}. Additionally to the standard holonomic constraints (5.1)$_3$, control constraints as given by

$$g_s = \bar{q} - \gamma(t) = 0 \tag{5.5}$$

are imposed to force the rigid body at hand to follow a prescribed motion $\gamma(t) = \begin{bmatrix} \gamma_x & \gamma_{d_i} \end{bmatrix}^T$. In Eq. (5.5), the function $\bar{q} : \mathcal{T} \mapsto \mathbb{R}^{12}$ has been defined as $\bar{q} = \begin{bmatrix} \bar{x} & d_i \end{bmatrix}.^T$ The servo-constraints (5.5) of course must not violate the holonomic constraints (5.1)$_3$. The differential part of system (5.1) along with the control constraint (5.5) gives rise to a purely algebraic system of equations given by

$$\bar{f}(t) = D\partial^2_t \gamma(t) + F\gamma(t) - \bar{G} \,. \tag{5.6}$$

Herein, the coefficients D, F and $\bar{G}$ has been defined as

$$D = \begin{bmatrix} M & 0 \\ 0 & E_{ij} \end{bmatrix} \quad , \quad F = \begin{bmatrix} 0 & 0 \\ 0 & \Lambda_{ij} \end{bmatrix} \quad , \quad \bar{G} = \begin{bmatrix} G \\ 0 \end{bmatrix} \,. \tag{5.7}$$

Eq. (5.6) determines the external (director-) force $\bar{f}(t)$ that is conjugate to the prescribed trajectory γ. Once $\bar{f}(t)$ has been computed, the $k \in \mathbb{N}$ contact forces $n^k_c(t)$ for the k strings at $s_k = 1$ can be easily computed by knowing the position of the contact point of the string at the rigid body. By defining the components Q_{kl} of the inverse of the positive definite matrix $\mathcal{H} \in \mathbb{R}^{4d,4d}$ introduced in (5.3), a linear relation of the contact forces to the (director-) forces is obtained as

$$n^k_c(t) = Q_{kl} \bar{f}_l(t). \tag{5.8}$$

Once the forces $n^k_c(t)$ have been calculated, each string can be solved separately by inserting the forces into the corresponding *Neumann* boundary condition (1.6) of the initial boundary value problem established in Sec. 1.1. This cascade-like solution strategy for the cooperative control problem enables a parallelisation of the computation of the inverse dynamics of the attached strings. The applicability of this approach is demonstrated in Ex. 5.3.

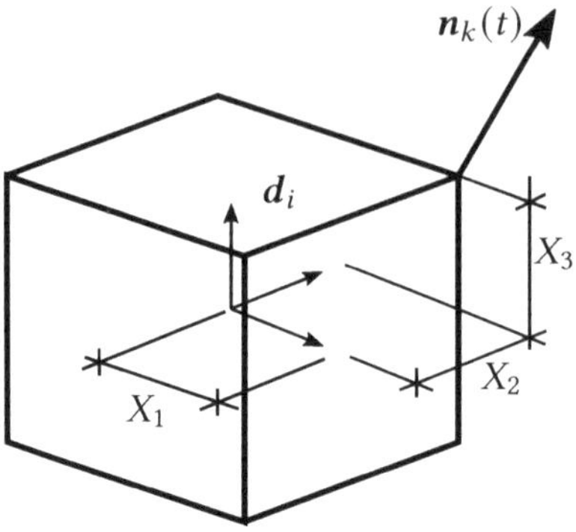

Figure 5.2.: Force $\boldsymbol{n}_k$ acting on a rigid body at point $\boldsymbol{P} = \boldsymbol{x} + X_i \boldsymbol{d}_i$.

Remark 5.1 (Lagrange multipliers). *The actuating forces depend on the Lagrange multipliers Λ_{ij}, i.e. an unique solution for the actuation of the rigid body requires 6 independent components $\bar{\lambda}_k : \mathcal{T} \mapsto \mathbb{R}$ for $k \in \{1, \cdots, 6\} \subset \mathbb{N}$ of Λ_{ij} to be partly specified.*

$$\lambda_k(t) = \gamma_{\lambda_k}(t)$$

Essentially this amounts to partly specifying the stresses within the rigid body (cf. [55]).

Multibody system. For the purpose of verification, the flexible multibody system is introduced for the direct dynamics problem. Therefore, consulting (3.4) along with (3.5) and (3.7) for the string and (5.1) for the rigid body, the motion of the overall system is governed by a differential-algebraic system of equations given by

$$\begin{aligned}
\boldsymbol{0} &= \boldsymbol{R}(\mathrm{D}_t^2 \boldsymbol{q}, \boldsymbol{q}, t) - (\partial_{\boldsymbol{q}} \boldsymbol{g}_k(\boldsymbol{q}))^T \mu \\
\boldsymbol{0} &= \boldsymbol{g}_k = \boldsymbol{r}_{p+1} - (\bar{\boldsymbol{x}} + X_i \boldsymbol{d}_i) \\
0 &= g_c = \boldsymbol{d}_i \cdot \boldsymbol{d}_j - \delta_{ij}
\end{aligned}$$

where

$$\boldsymbol{R}(\mathrm{D}_t^2 \boldsymbol{q}, \boldsymbol{q}, \bar{\lambda}, t) = \begin{bmatrix} \hat{\boldsymbol{M}} \, \mathrm{D}_t^2 \hat{\boldsymbol{q}} + \boldsymbol{f}_i^{int} - \hat{\boldsymbol{G}} \\ \boldsymbol{D} \, \mathrm{D}_t^2 \bar{\boldsymbol{q}} + \boldsymbol{F} \bar{\boldsymbol{q}} - \bar{\boldsymbol{G}} \end{bmatrix} .$$

Therein the solution $\boldsymbol{q}$ is defined by $\boldsymbol{q} = \begin{bmatrix} \hat{\boldsymbol{q}}^T & \bar{\boldsymbol{q}}^T \end{bmatrix}^T$. Note that the Lagrange multipliers $\mu : T \mapsto \mathbb{R}^d$ in conjunction with the constraint Jacobian $\partial_{\boldsymbol{q}} \boldsymbol{g}_k(\boldsymbol{q})$ can be identified as the contact force acting among the string and the rigid body

$$\mu \partial_{\boldsymbol{q}} \boldsymbol{g}(q) = \begin{bmatrix} 0 & \cdots & \mu & -\mu & -\mu X_1 & -\mu X_2 & -\mu X_3 \end{bmatrix}^T \tag{5.9}$$

See Ex. 5.2 for an application of this model to a simulation of a free swinging rigid body attached to a single elastic string.

Example 5.2 (Direct dynamics problem - pendulum). *To verify the numerical implementation of the procedure outlined in this section, an example for the forward dynamics of a rigid body suspended by a flexible string is considered. Therefore, a cube with edge length* $a = 2$ *and total mass* $M = 0.1$ *is attached to an elastic string characterized by its Young's Modulus* $E = 1$*, density* $\rho_0 = 1$*, cross-sectional area* $A = 1$ *and length* $L = 1$*. The velocity* $v(t)$ *of the rigid cube at time* $t = 0.0$ s *is assumed to be* $v_0 = \begin{bmatrix} 2 & -5 & 0 \end{bmatrix}^T$*. In Fig. 5.3 snapshots of the swinging rigid body are depicted.*

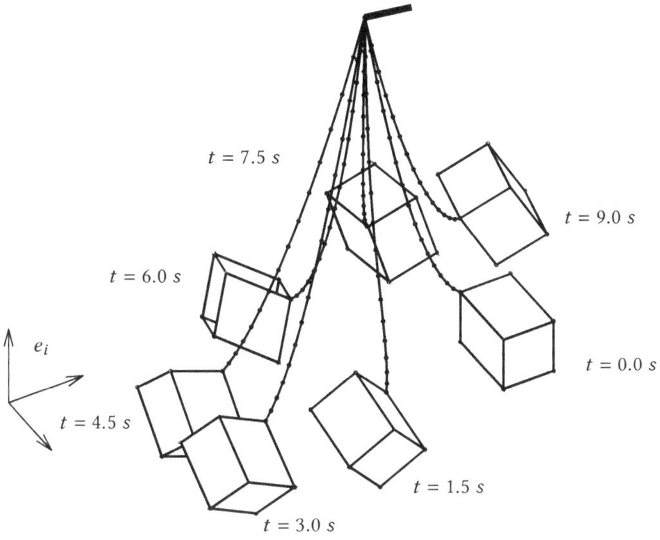

Figure 5.3.: Free oscillation of a rigid cube with edge lenth $a = 2$ and total mass $\bar{M} = 0.1$ connected to a flexible rope with $\rho = 1$ and $E = 2$.

Example 5.3 (Inverse dynamics problem - rotation and translation). *To verify the presented approach to the inverse dynamics problem under consideration, the following scenario is given. A rigid cube with edge length* $a = 2$ *and mass* $M = 1$ *is supposed to accomplish a rest-to-rest maneuver. For this a translation of the cube from point* $P_0 = (0, 0, 0)$ *to point* $P_f = (2, 2, 2)$ *along a straight line together with a simultaneous rotation of* π *around the z-axis is planned. The maneuver is intended to start at* $t_0 = 1.0$ s *and end at* $t_f = 9.0$ s*. The motion of the rigid body described in terms of a Cosserat point subjected to geometric constraints can then be prescribed together with*

$$\varphi = -\frac{\pi}{2}\left[\cos\left(\frac{\pi}{2}(\sin\theta + 1)\right) - 1\right] \qquad \textit{for} \qquad \theta = \pi\left(\frac{t - t_0}{t_0 - t_f} - \frac{1}{2}\right) \tag{5.10}$$

through the trajectory as given by

$$\gamma_{rb} = \begin{bmatrix} 0\\0\\0\\1\\0\\0\\0\\1\\0\\0\\0\\1 \end{bmatrix} \quad \forall\, t < t_0, \quad \gamma_{rb} = \begin{bmatrix} -\cos\varphi + 1\\ -\cos\varphi + 1\\ -\cos\varphi + 1\\ \cos\varphi\\ \sin\varphi\\ 0\\ -\sin\varphi\\ \cos\varphi\\ 0\\0\\0\\1 \end{bmatrix} \quad \forall\, t \in T_m = [t_0, t_f], \quad \gamma_{rb} = \begin{bmatrix} 2\\2\\2\\-1\\0\\0\\0\\-1\\0\\0\\0\\1 \end{bmatrix} \quad \forall\, t > t_f \tag{5.11}$$

Thereby, four elastic ropes with mass density $\rho = 1$ and Young's Modulus $E = 1$, are used to actuate the rigid body for which the motion is prescribed. Regarding the k predefined contact points for $i \in \{1, 2, 3\}$

$$\boldsymbol{P}_k = \bar{\boldsymbol{x}} + X_i^k \boldsymbol{d}_i$$

and following Sec. 5.1 the contact forces needed to achieve the desired motion can then be computed using (5.8) with

$$\mathcal{H} = \begin{bmatrix} \boldsymbol{I} & \boldsymbol{I} & \boldsymbol{I} & \boldsymbol{I} \\ \boldsymbol{I} & -\boldsymbol{I} & \boldsymbol{I} & -\boldsymbol{I} \\ \boldsymbol{I} & -\boldsymbol{I} & -\boldsymbol{I} & \boldsymbol{I} \\ -\boldsymbol{I} & \boldsymbol{I} & \boldsymbol{I} & \boldsymbol{I} \end{bmatrix}$$

where $\boldsymbol{I} \in \mathbb{R}^{d,d}$ is the identity matrix. Note that, following Rem. 5.1, the Lagrange multipliers cannot be left undefined. For the given maneuver, a uniaxial tension within the rigid body is choosen:

$$\bar{\boldsymbol{\lambda}} = \begin{bmatrix} 0 & 0 & \bar{M}g & 0 & 0 & 0 \end{bmatrix}^T \tag{5.12}$$

In Fig. 5.4 components of the k contact forces are shown. The forces acting on the upper end of the k ropes at $s_k = 0$ can be calculated such that the rigid body at hand follows the prescribed trajectory. The numerical solution is shown in Fig. 5.5.

To verify the outlined method, the forces $\boldsymbol{f}_k(t)$ can be inserted into the flexible multibody system presented in Sec. 5.1. In Fig. 5.6, snapshots of a forward simulation of the flexible multibody system at hand actuated with the forces $\boldsymbol{f}_k(t)$ acting at the upper ends of the k ropes computed numerically using the approach presented above are shown.

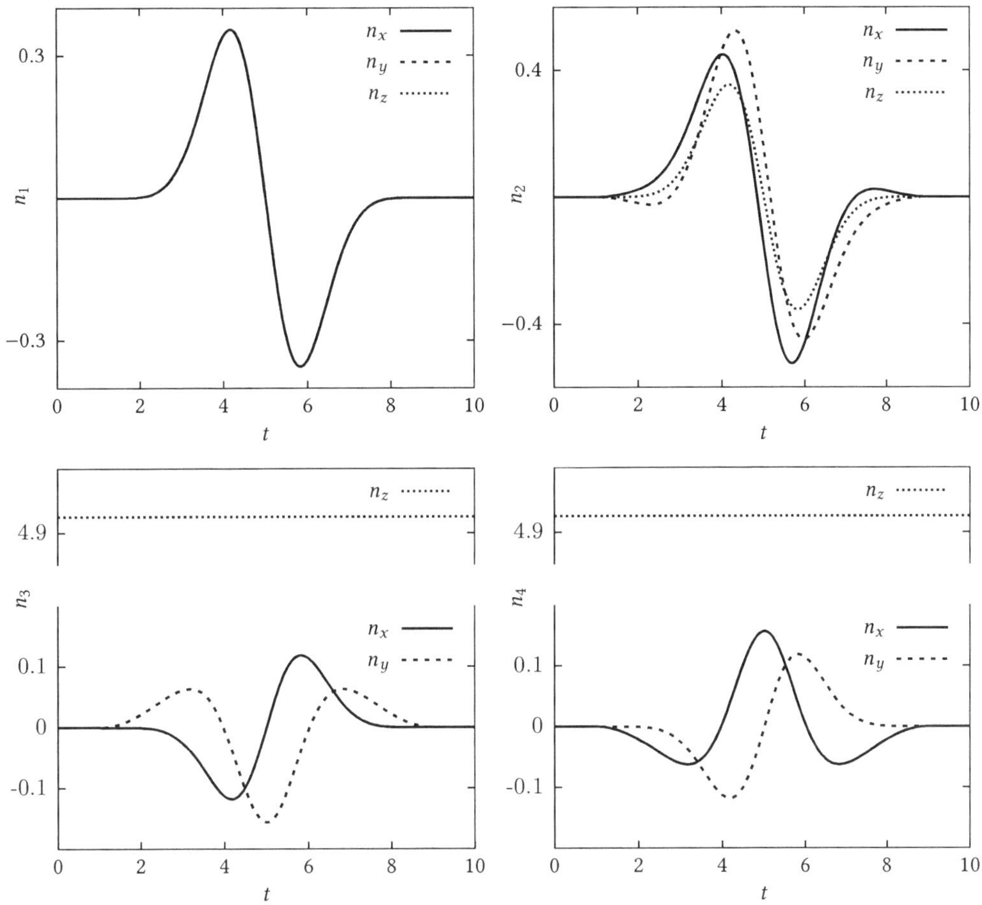

Figure 5.4.: Components of the contact forces $n_k(t)$ such that the rigid body follows the prescribed translation of the rigid body from point $P_0 = (0, 0, 0)$ to point $P_f = (2, 2, 2)$ accompanied by a simultaneous rotation π.

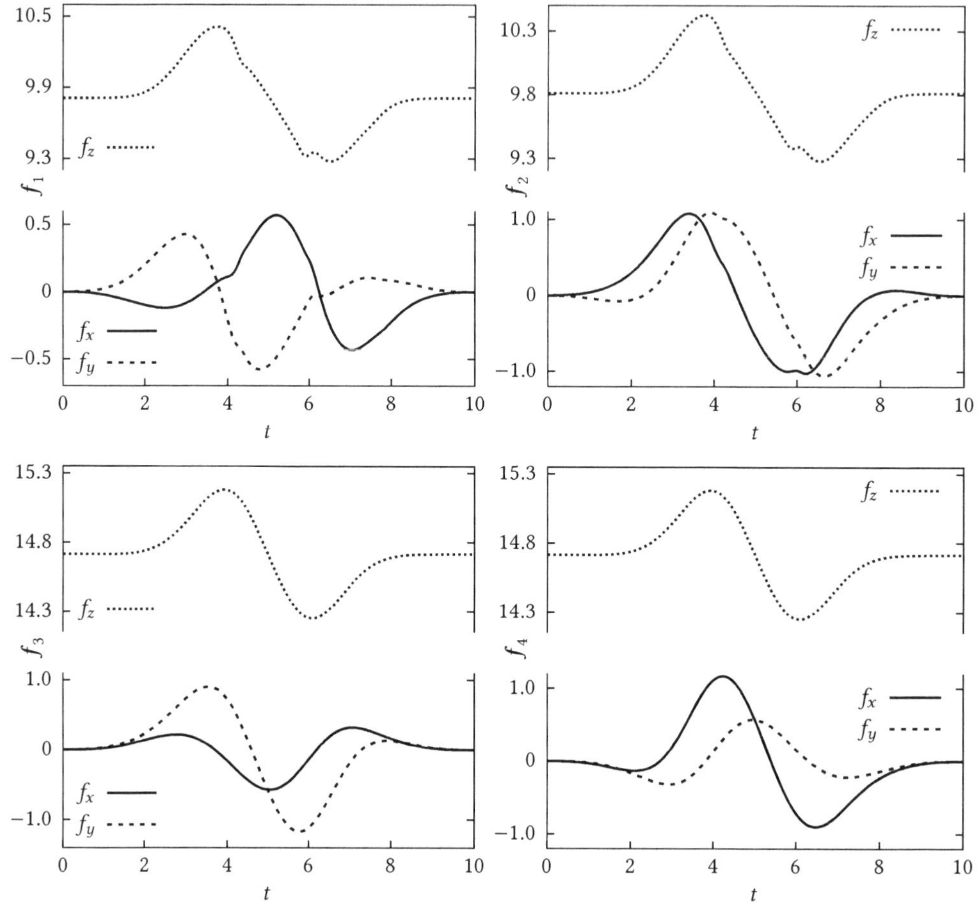

Figure 5.5.: Components of the actuating forces $\boldsymbol{f}_k(t)$ for $k \in \{1, 2, 3, 4\}$ acting on the upper end of the ropes at $s_k = 0$ accounting for the prescribed motion of the rigid body.

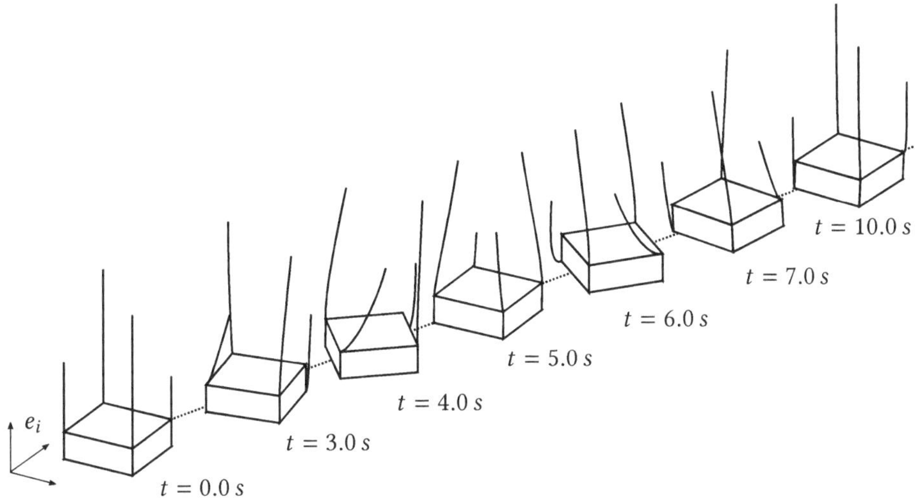

Figure 5.6.: Snapshots of a rigid cube with edge length $a = 2$ and total mass $\bar{M} = 1$ actuated through four flexible ropes with $\rho = 1$ and $E = 2$ following a straight line from $\boldsymbol{P}_0 = (0, 0, 0)$ to $\boldsymbol{P}_f = (2, 2, 2)$ while rotating simultaneously by the prescribed angle π.

5.2. Rotational arm model

A rotational arm model consisting of a chain of *three* arms is investigated subsequently. Particularly, a flexible arm $\mathcal{B}_3$ is attached to a rigid arm $\mathcal{B}_2$ in point C, which in turn is connected to another rigid arm $\mathcal{B}_1$ in point B. Body $\mathcal{B}_1$ in turn is attached to an inertial frame in point A. See Fig. 5.7 for an illustration thereof. The bodies are assumed to have density $\rho_i : \mathcal{B}_i \mapsto \mathbb{R}$ and total mass $M_i = \int_{\mathcal{B}_i} \rho_i \,\mathrm{dV}$ for $i \in \{1, 2, 3\}$. The task is to find the torques $\boldsymbol{m}_A$, $\boldsymbol{m}_B$ and $\boldsymbol{m}_C$ that need to be applied to the joints A, B and C, respectively, such that the end-effector P follows a prescribed motion $\boldsymbol{\gamma}(t)$ given by

$$\boldsymbol{\gamma} = \begin{bmatrix} 1 \\ 0 \\ 0 \end{bmatrix} \forall t < t_0 \qquad \boldsymbol{\gamma} = \begin{bmatrix} 1 - \cos(\Theta \cdot y(t)) \\ \sin(\Theta \cdot y(t)) \\ \Theta \cdot y(t) \end{bmatrix} \forall t \in [t_0, t_0 + T] \qquad \boldsymbol{\gamma} = \begin{bmatrix} 0 \\ 1 \\ \Theta \end{bmatrix} \forall t > t_f \,. \quad (5.13)$$

In Eq. (5.13) the angle of the prescribed overall rotation $\Theta \in [0, 2\pi]$ and the function

$$y(t) = 1 - \frac{1}{2}\left(\cos\left(\frac{\pi}{2}\left(\sin\left(\pi\frac{t - t_0}{T} - \frac{\pi}{2}\right) + 1\right)\right) + 1\right)$$

has been introduced.

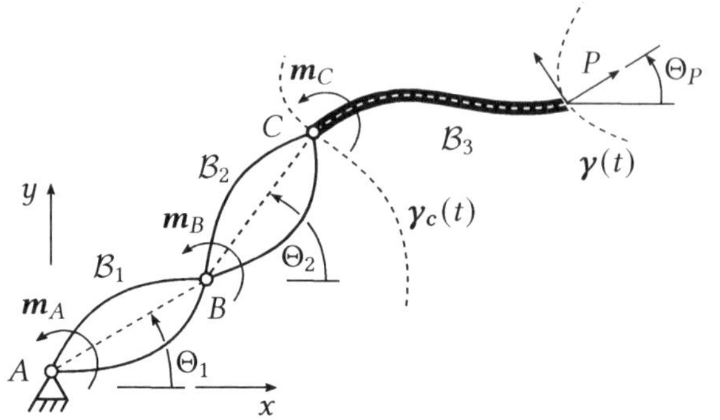

Figure 5.7.: Rotational arm model. Geometrically exact beam in context of flexible multibody system (2d). The coordinates (x_P, y_P) as well as angle Θ_P of the end-effector P denote the system outputs. The torques $(\boldsymbol{m}_A, \boldsymbol{m}_B, \boldsymbol{m}_C)$ in joints (A, B, C) are the conjugate system inputs.

In principal, the same cascade-like procedure, that could have been developed in Sec. 5.1 in context of the cooperative control problem, can be applied to the inverse dynamics of the flexible rotational arm model at hand. The strategy reads as follows.

The flexible arm is modelled as a *Cosserat rod* aligning with the framework introduced in Sec. 1.1 and Sec. 2.2, analyzed in Ex. 4.5 as well as Ex. 4.4 and solved numerically in Ex. 4.30 and Ex. 4.29. The proposed numerical schemes based on a simultaneous space-time integration, do not only provide a solution for the actuating force searched for. They also give rise to the overall motion, including the trajectory of the actuating joint C as given by

$$\boldsymbol{\gamma}_C(t) = \begin{bmatrix} x_C(t) \\ y_C(t) \end{bmatrix} . \tag{5.14}$$

In case of describing the kinematics of the rigid part of the system in Cartesian coordinates, i.e. in a redundant configuration space, the trajectory $\boldsymbol{\gamma}_C$ can directly be inserted into the differential algebraic system of equations. Similarly to Sec. 5.1, the external applied (director) forces need then be determined.

Alternatively, minimal coordinates can be used to describe the kinematics of the rigid multibody system. Then we need to find coordinates Θ_1 and Θ_2, aligning with the computed trajectory of the sub-systems end-effector (joint C), i.e.

$$\boldsymbol{\gamma}_C(t) = \begin{bmatrix} x_C(t) \\ y_C(t) \end{bmatrix} = \begin{bmatrix} \cos\Theta(t) + \cos\Theta_2(t) \\ \sin\Theta_1(t) + \sin\Theta_2(t) \end{bmatrix} = \mathcal{F}(\Theta_1, \Theta_2) . \tag{5.15}$$

Obviously, the *inverse kinematics* (5.15) are not uniquely solvable[1] for the minimal coordinates $\Theta_1(t)$ and $\Theta_2(t)$. We therefore solve (5.15) in terms of a least-square ansatz:

$$\left\|\mathcal{F}(\Theta_1, \Theta_2) - \boldsymbol{\gamma}_C\right\|_2^2 \to Min.$$

[1] By considering a feasible configuration $\bar{\Theta}_1 = \Theta_2$ and $\bar{\Theta}_2 = \Theta_1$ such that $\bar{\gamma}(t) = \gamma(t)$ but $\bar{x}_1(t) \neq x_1(t)$ and $\bar{x}_2(t) \neq x_2(t)$ proves the singularity of (5.15)

Once the motion of the remaining rigid part of the model has been calculated, the actuating torques, can be determined uniquely by evaluating the governing balance equations of linear and angular momenta of the rigid part of the model.

Since the contact force $\boldsymbol{n}_C$ acting at joint C is determined by the inverse computation of the flexible beam $\mathcal{B}_3$, the contact force in joint B can be obtained, by using the balance of linear momentum of the rigid body $\mathcal{B}_2$, as

$$\boldsymbol{n}_B = \mathrm{D}_t \boldsymbol{P}_2 - \boldsymbol{G}_1 - \boldsymbol{n}_C \ . \tag{5.16}$$

In Eq. (5.16), the linear momentum of body $\mathcal{B}_2$ has been introduced as $\mathrm{D}_t \boldsymbol{P}_2 = M_2 \boldsymbol{a}_2$. Once the contact force $\boldsymbol{n}_B$ is determined, the balance of angular momentum of the rigid body $\mathcal{B}_2$ provides the contact torque $\boldsymbol{m}_B$ acting at joint B

$$\boldsymbol{m}_C - \boldsymbol{m}_B + (\boldsymbol{r}_2 \times \boldsymbol{G}_2) + (\boldsymbol{r}_B \times \boldsymbol{n}_B) + (\boldsymbol{r}_C \times \boldsymbol{n}_C) = \mathrm{D}_t \boldsymbol{L}_{A_2} \ . \tag{5.17}$$

Then the torque $\boldsymbol{m}_A$ acting at joint A can be calculated by establishing the balance of angular momentum of body $\mathcal{B}_1$

$$\boldsymbol{m}_A + \boldsymbol{m}_B + (\boldsymbol{r}_1 \times \boldsymbol{G}_1) + (\boldsymbol{r}_B \times \boldsymbol{n}_B) = \mathrm{D}_t \boldsymbol{L}_{A_1} \ . \tag{5.18}$$

In Eq. (5.17) and Eq. (5.18), the gravitational forces $\boldsymbol{G}_1$ and $\boldsymbol{G}_2$, acting at the center of gravity of body $\mathcal{B}_1$ and $\mathcal{B}_2$, respectively, have been introduced. The torques $\boldsymbol{m}_A$ and $\boldsymbol{m}_B$ applied to joints A and B, respectively, can be identified as differentially flat inputs corresponding to the differentially fat output, by means of the trajectory of the end-effector of the fully actuated rigid multibody system. In Fig. 5.8 the inverse computed actuating torques $\boldsymbol{m}_A$, $\boldsymbol{m}_B$ and $\boldsymbol{m}_C$ are depicted.

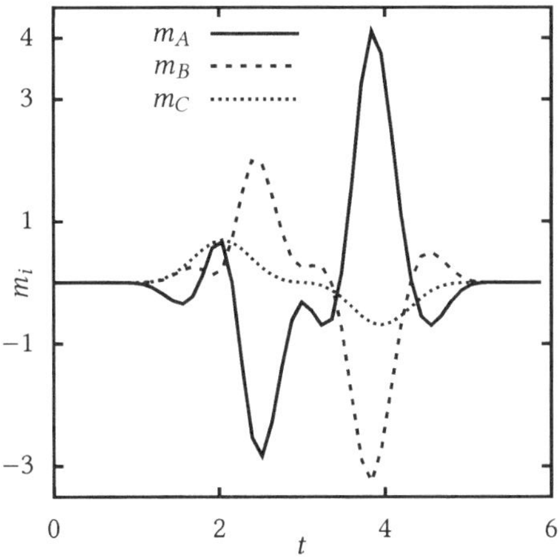

Figure 5.8.: Actuating torques m_i for $i \in \{A, B, C\}$, applied to joints A, B and C such that the end-effector captures the prescribed motion.

6. Conclusion

6.1. Summary

The present work has focused on the inverse dynamics of underactuated flexible mechanical systems undergoing large deformations. Therefore, it could be demonstrated that this specific problem is governed by quasi-linear second-order partial differential equations subjected to time-varying *Dirichlet* boundary conditions that are enforced by spatially disjunct *Neumann* boundary conditions. Various mechanical systems including slender structures such as strings and beams as well as general *three*-dimensional continua have been presented and proved to align with this specific initial boundary value problem.

Then, a brief survey on common solution strategies relying on a sequential space-time discretization has been given. Therefore, it could be elaborated that such methods are leading either to (unstable) internal dynamics or high differentiation index' along with high demands on the smoothness of the prescribed constraint function. Both hinder a numerical stable integration of the semi-discrete equations. We were able to demonstrate, that these issues are caused by the non-standard construction of the constraint realization in conjunction with the applied sequential space-time discretization.

At this point, an in-depth analysis of the initial boundary value problem at hand was undertaken. In that process, characteristic manifolds along which information flows could be identified and the hyperbolic structure of the governing partial differential equations has been revealed. Enlightning these wave phenomena has provided precious insights into the governing equations and has paved the way to develop novel numerical methods that are capable of solving the inverse dynamics of (infinite-) dimensional systems.

Therefore, a method that is based on a geometrical interpretation of the underlying partial differential equations could be introduced. It has been demonstrated concisely that once characteristic manifolds have been found, the partial differential equations at hand can be transformed equivalently into a system of ordinary differential equations along these manifolds. The succesful implementation of this method has motivated to develop a novel Galerkin-based space-time finite element method. This space-time finite element approach relies on continuous test functions. This is in sharp contrast to previously developed space-time finite element methods that rely on discontinuous test functions thus making possible to divide the space-time domain into a sequence of space-time slabs eventually recovering the time-marching format commonly applied in structural dynamics.

Then, these novel numerical methods could be adopted to the feed-forward control problem of flexible multibody systems. Therefore, a rigid body that is cooperatively controlled through several elastic strings and a rotational arm consisting of rigid and flexible arms has been considered.

Selected representative numerical examples have been presented demonstrating the capabilities of the newly devised space-time integration methods.

6.2. Outlook

This work has answered some important questions but also raised new ones. Some of them are briefly summarized below.

The simultaneous space-time integration methods proved to be the key to succesively solve the inverse dynamics problem of infinite-dimensional systems. Thereby it turned out, that the method of characteristics is very well adapted to the governing equations. On the other hand the space-time finite element method has proved to be versatile applicable due to its inherent simplicity. However, its potential has not been fully exploited yet. For instance, the possibility of using higher order elements as well as using unstructured meshes along with an error-based adaptive mesh refinement have not been implemented so far. Moreover a combination of the method of characteristics and the space-time finite element method might be conceivable by orienting the finite elements along characteristic lines. Such a characteristic Galerkin method may combine the accuracy of the method of characteristics and the simplicity of the space-time finite element method.

Besides the spatially continuous systems, spatially finite-dimensional systems constitute another important class of inverse dynamics problems. Such systems are governed inherently by ordinary differential equations and hence elude from the framework proposed for spatially infinite-dimensional systems. Since the inverse dynamics of spatially discrete sytems such as cranes or manipulators with passive joints are highly relevant, two global Galerkin-based methods could have been developed in Rem. 4.21 and Rem. 4.22. One is derived directly from the continuous space-time finite element method presented in Sec. 4.2 and the other one is based on discontinuous approximations of the test- and trial functions along with additional flux terms propagating backwards in time. These fluxes are accounting for the causality of time-dependent physical systems. Further analysis seams to be promising.

Throughout this work, the motion of mechanical systems has partly been specified. But for some cases it might be beneficial to find trajectories connecting two prescibed points that are optimal in terms of causing a minimal effort of the actuation. Questions on the existence and uniqueness of a trajectory quickly arise, hindering a stable numerical solution. Therefore, adopting the newly gained insights, regarding the property of physical time-dependent delay systems and its connection to the property of differential flatness of spatially discrete systems, to the theory of optimal control, might be convenient to find feasible trajectories.

Publications and talks

Publications

- Ströhle T, Betsch P, and Schiebl M. "Galerkin-based mixed time finite elements for structural dynamics". In: *PAMM Proceedings of Applied Mathematics and Mechanics.* 18(1), 2018.
- Ströhle T and Betsch P. "Inverse dynamics of underactuated flexible mechanical systems". In: *PAMM Proceedings of Applied Mathematics and Mechanics.* 19(1), 2019.
- Ströhle T and Betsch P. "Solution Techniques for Problems of Inverse Dynamics of Flexible Underactuated Systems". In: *Computational Methods in Applied Sciences, ECCOMAS Thematic Conference on Multibody Dynamics 2019.* Ed. by Kecskeméthy A and Geu Flores F. Vol. 53. Cham: Springer International Publishing, 2020.
- Ströhle T and Betsch P. "Controlling Nonlinear Elastic Systems in Structural Dynamics". In: *14th World Congress on Computational Mechanics (WCCM) and European Congress on Computational Methods in Applied Sciences and Engineering (ECCOMAS).* 2021.
- Ströhle T and Betsch P. "Simultaneous Space-Time Discretization for Controlling the Motion of Rigid Bodies Actuated Through Elastic Ropes". In: *European Congress on Computational Methods in Applied Sciences and Engineering (ECCOMAS) - Thematic Conference on Multibody Dynamics.* 2021.
- Ströhle T and Betsch P. "A simultaneous space-time discretization approach to the inverse dynamics of geometrically exact strings". In: *International Journal for Numerical Methods in Engineering.* 123(11): 2573–2609, 2022.
- Ströhle T and Betsch P. "Inverse dynamics of geometrically exact beams". In: *European Congress on Computational Methods in Applied Sciences and Engineering (ECCOMAS).* 2022.

Conference talks

- Ströhle T and Betsch P. *Controlling nonlinear elastic systems in structural dynamics.* 14th World Congress on Computational Mechanics (WCCM) and European Congress on Computational Methods in Applied Sciences and Engineering (ECCOMAS). Paris, France, Virtual Congress, January 11-15, 2021.

- Ströhle T and Betsch P. *Controlling the equations of motion for the classical form of Cosserat rods.* International Union for Theoretical and Applied Mechanics (IUTAM): Symposium on Optimal Design and Control of Multibody Systems. Hamburg, Germany, July 18-21, 2022.
- Ströhle T and Betsch P. *Galerkin space-time discretization approach to the inverse dynamics of flexible multibody systems.* Centre International de Rencontres Mathématiques (CIRM): Conference on Energy-Based Modeling, Simulation, and Control of Complex Constrained Multiphysical Systems. Luminy, Marseille, France, April 18-22, 2022.
- Ströhle T and Betsch P. *Inverse Dynamics of Geometrically Exact Beams.* 8th European Congress on Computational Methods in Applied Sciences and Engineering (ECCOMAS). Oslo, Norway, June 5-9, 2022.
- Ströhle T and Betsch P. *Inverse Dynamics of Nonlinear Elastic Ropes and the Method of Characteristics.* 91st Annual Meeting of the International Association of Applied Mathematics and Mechanics (GAMM). Kassel, Germany, Virtual Congress, March 15-19, 2021.
- Ströhle T and Betsch P. *Inverse dynamics of underactuated flexible mechanical systems.* 90th Annual Meeting of the International Association of Applied Mathematics and Mechanics (GAMM). Vienna, Austria, February 18-22, 2019.
- Ströhle T and Betsch P. *Simultaneous Space-Time Discretization for Controlling the Motion of Rigid Bodies Actuated Through Elastic Ropes.* 10th European Congress on Computational Methods in Applied Sciences and Engineering (ECCOMAS),Thematic Conference on Multibody Dynamics. Budapest, Hungary, Virtual Congress, December 12-15, 2021.
- Ströhle T and Betsch P. *Solution techniques for problems of inverse dynamics of flexible underactuated systems.* 9th European Congress on Computational Methods in Applied Sciences and Engineering (ECCOMAS), Thematic Conference on Multibody Dynamics. Duisburg, Germany, July 15-18, 2019.
- Ströhle T and Betsch P. *Space-time integration methods for the inverse dynamics simulation of flexible mechanical systems.* 92nd Annual Meeting of the International Association of Applied Mathematics and Mechanics (GAMM). Aachen, Germany, August 15-19, 2022.
- Ströhle T, Betsch P, and Schiebl M. *Galerkin-based mixed time finite elements for structural dynamics.* 89th Annual Meeting of the International Association of Applied Mathematics and Mechanics (GAMM). Munich, Germany, March 19-23, 2018.

Further academic activities

- Ströhle T and Drücker S. *Organization of the Young Researcher Minisymposium: Dynamic Inversion and Control of Mechanical Underactuated Systems.* 92nd Annual Meeting of the International Association of Applied Mathematics and Mechanics (GAMM). Aachen, Germany, August 15-19, 2022.

List of Figures

Bibliography

[1] Abbott M. *An Introduction to the Method of Characteristics.* Elsevier, NY, 1966.

[2] Achenbach J. *Wave propagation in elastic solids.* 1st. North-Holland Elsevier, 1973.

[3] Ainsworth M. "Essential boundary conditions and multi-point constraints in finite element analysis". In: *Comput. Methods Appl. Mech. Engrg.* 190: 6323–6339, 2001.

[4] Altmann R, Betsch P, and Yang Y. "Index reduction by minimal extension for the inverse dynamics simulation of cranes". In: *Multibody System Dynamics.* 36(3): 295–321, 2016.

[5] Altmann R and Heiland J. "Simulation of multibody systems with servo constraints through optimal control". In: *Multibody System Dynamics.* 40: 75–98, 2017.

[6] Anderson M and Kimn JH. "A numerical approach to space-time finite elements for the wave equation". In: *Journal of Computational Physics.* 226(1): 466–476, 2007.

[7] Antman S. *Nonlinear Problems of Elasticity.* 2nd. Springer-Verlag, 1995.

[8] Arens T, Hettlich F, Karpfinger C, Kockelkorn U, Lichtenegger K, and Stachel H. *Mathematik.* Spektrum Akademischer Verlag, 2011.

[9] Argyris J and Scharpf D. "Finite Elements in Time and Space". In: *Nuclear Engineering and Design.* 10: 456–464, 1969.

[10] Arnold M. "A Lie Algebra Approach to Lie Group Time Integration of Constrained Systems". In: *Structure-preserving Integrators in Nonlinear Structural Dynamics and Flexible Multibody Dynamics.* Ed. by Betsch P. Cham: Springer International Publishing, 2016, pp. 91–158.

[11] Arnold V. *Mathematical Methods of Classical Mechanics.* 2nd. Springer, New York, 1989.

[12] Arnold V. *Ordinary Differential Equations.* 1st. Springer Berlin, Heidelberg, 1992.

[13] Ascher U and Petzold L. *Computer methods for ordinary differential equations and differential-algebraic equations.* SIAM, 1998.

[14] Bastos G, Seifried R, and Brüls O. "Analysis of stable model inversion methods for constrained underactuated mechanical systems". In: *Mechanism and Machine Theory.* 111: 99–117, 2017.

[15] Baumgarte J. "Stabilization of constraints and integrals of motion in dynamical systems". In: *Computer Methods in Applied Mechanics and Engineering.* 1(1): 1–16, 1972.

[16] Behr M. "Simplex space-time meshes in finite element simulations". In: *Int. J. Numer. Meth. Fluids.* 57: 1421–1434, 9 2008.

[17] Berger T, Drücker S, Lanza L, Reis T, and Seifried R. "Tracking control for underactuated non-minimum phase multibody systems". In: *Nonlinear Dynamics.* 104: 3671–3699, 2021.

[18] Berger T, Otto S, Reis T, and Seifried R. "Combined open-loop and funnel control of underactuated multibody systems". In: *Nonlinear Dynamics.* 95(3): 1977–1998, 2019.

[19] Betsch P and Steinmann P. "Conservation Properties of a Time FE Method. Part I: Time-Stepping Schemes for N-Body Problems". In: *Int. J. Numer. Meth. Engng.* 49: 599–638, 2000.

[20] Betsch P and Steinmann P. "Frame-indifferent beam finite elements based upon the geometrically exact beam theory". In: *International Journal for Numerical Methods in Engineering.* 54(12): 1775–1788, 2002.

[21] Betsch P. "Energy-Momentum Integrators for Elastic Cosserat Points, Rigid Bodies, and Multibody Systems". In: *Structure-preserving Integrators in Nonlinear Structural Dynamics and Flexible Multibody Dynamics.* Ed. by Betsch P. Cham: Springer International Publishing, 2016, pp. 31–89.

[22] Blajer W. "Dynamics and control of mechanical systems in partly specified motion". In: *J. Franklin Inst.* 334B(3): 407–426, 1997.

[23] Blajer W. "The Use of Servo-Constraints in the Inverse Dynamics Analysis of Underactuated Multibody Systems". In: *J. Comput. Nonlinear Dynam.* 9(4): 041008/1–11, 2014.

[24] Blajer W and Kołodziejczyk K. "A Geometric Approach to Solving Problems of Control Constraints: Theory and a DAE Framework". In: *Multibody System Dynamics.* 11: 343–364, 2004.

[25] Blajer W, Seifried R, and Kołodziejczyk K. "Diversity of Servo-Constraint Problems for Underactuated Mechanical Systems: A Case Study Illustration". In: *Solid State Phenomena.* 198: 473–482, 2013.

[26] Blajer W, Seifried R, and Kołodziejczyk K. "Servo-constraint realization for underactuated mechanical systems". In: *Archive of Applied Mechanics.* 85: 1191–1207, 9-10 2015.

[27] Bolza O. *Vorlesungen über Variationsrechnung.* Leipzig und Berlin, 1909.

[28] Campbell S. "High-Index Differential Algebraic Equations". In: *Mech. Struct. & Mach.* 23(2): 199–222, 1995.

[29] Chorin A and Marsden J. *A Mathematical Introduction to Fluid Mechanics.* Vol. 3. Springer, 2000.

[30] Cosserat E and Cosserat F. *Théorie des Corps déformables. Librairie Scientifique.* A. Hermann et Fils, 1909.

[31] Courant R and Friedrichs K. *Supersonic Flow and Shock Waves.* Interscience, 1948.

[32] Courant R and Hilbert D. *Methoden der mathematischen Physik I.* Springer-Verlag, 1924.

[33] Courant R and Hilbert D. *Methoden der mathematischen Physik II.* Springer-Verlag, 1937.

[34] Crisfield M. *Non-linear Finite Element Analysis of Solids and Structures. Volume 2: Advanced Topics.* John Wiley & Sons, 1997.

[35] Delfour M, Hager W, and Trochu F. "Diskontinuous Galerkin Methods for ordinary differential equations". In: *Mathematics of Computation.* 36(154): 455–473, 1981.

[36] Eriksson K, Estep D, Hansbo P, and Johnson C. *Computational Differential Equations.* Cambridge University Press, 1996.

[37] Evans L. *Partial differential equations.* Providence, R.I.: American Mathematical Society, 2010.

[38] Fliess M, Lévine J, Martin P, and Rouchon P. "Flatness and defect of non-linear systems: introductory theory and examples". In: *International Journal of Control.* 61: 2046–2051, 1995.

[39] Fliess M, Mounier H, Rouchon P, and Rudolph J. "Controllability and motion planning for linear delay systems with an application to a flexible rod". In: *Proceedings of 1995 34th IEEE Conference on Decision and Control.* 2: 2046–2051, 1995.

[40] Gear C. "Differential-Algebraic Equation Index Transformations". In: *SIAM Journal on Scientific and Statistical Computing.* 9(1): 39–47, 1988.

[41] Gear C. "Differential Algebraic Equations, Indices, and Integral Algebraic Equations". In: *SIAM Journal on Numerical Analysis.* 27(6): 1527–1534, 1990.

[42] Gear C, Leimkuhler B, and Gupta G. "Automatic integration of Euler-Lagrange equations with constraints". In: *Journal of Computational and Applied Mathematics.* 12-13: 77–90, 1985.

[43] Géradin M and Rixen D. *Mechanical Vibrations.* Wiley, 2015.

[44] Goldstein H, Pool C, and Safko J. *Classical Mechanics.* 3rd. Addison Wesley, 2001.

[45] Hadamard J. "Sur les Problèmes aux Dérivées Partielles et Leur Signification Physique". In: *Princeton University Bulletin.* 13: 49–52, 1902.

[46] Hadamard J. *Lecons sur le calcul de variations I.* Paris, 1910.

[47] Hesch C, Schuß S, Dittmann M, Eugster S, Favino M, and Krause R. "Variational Space-Time Elements for Large Systems". In: *Comput. Methods Appl. Mech. Engrg.* 326: 541–572, 2017.

[48] Hilbert D. "Zur Variationsrechnung". In: *Mathematische Annalen.* 62: 351–370, 1906.

[49] Hilbert D. "Über den Begriff der Klasse von Differentialgleichungen". In: *Mathematische Annalen.* 73: 95–108, 1912.

[50] Holzapfel G. *Nonlinear Solid Mechanics: A Continuum Approach for Engineering.* Wiley, 2000.

[51] Hughes T. *The Finite Element Method.* Dover Publications, 2000.

[52] Hughes T and Hulbert G. "Space-Time Finite Element Methods for Elastodynamics: Formulations and Error Estimates". In: *Comput. Methods Appl. Mech. Engrg.* 66: 339–363, 1988.

[53] Idesman A. "Solution of linear elastodynamics problems with space-time finite elements on structured and unstructured meshes". In: *Comput. Methods Appl. Mech. Engrg.* 196(9-12): 1787–1815, 2007.

[54] Idesman A, Niekamp R, and Stein E. "Continuous and discontinuous Galerkin methods with finite elements in space and time for parallel computing of viscoelastic deformation". In: *Comput. Methods Appl. Mech. Engrg.* 190(8-10): 1049–1063, 2000.

[55] Irscheid A, Konz M, and Rudolph J. "A Flatness-Based Approach to the Control of Distributed Parameter Systems Applied to Load Transportation with Heavy Ropes". In: *Advanced Control Techniques in Complex Engineering Systems: Theory and Applications.* Ed. by Kondratenko Y, Chikrii A, Gubarev V, and Kacprzyk J. Cham: Springer International Publishing, 2019, pp. 279–294.

[56] Isidori A. *Nonlinear Control Systems.* 3rd. Springer London, 1995.

[57] John F. *Partial Differential Equations.* Springer, 1982.

[58] Kirsch A. *An Introduction to the Mathematical Theory of Inverse Problems.* Springer, New York, NY, 2021.

[59] Klein F and Sommerfeld A. *Über die Theorie des Kreisels : in 4 Heften.* 1965.

[60] Knüppel T and Woittennek F. "Control Design for Quasi-Linear Hyperbolic Systems With an Application to the Heavy Rope". In: *IEEE Transactions on Automatic Control.* 60(1): 5–18, 2015.

[61] Kolsky H. *Stress Waves in Solids.* Oxford University Press, 1953.

[62] Krenk S. *Non-linear Modeling and Analysis of Solids and Structures.* Cambridge University Press, 2009.

[63] Kunkel P and Mehrmann V. *Differential-Algebraic Equations - Analysis and Numerical Solution.* EMS Textbooks in Mathematics, 2006.

[64] Lam S. "On Lagrangian dynamics and its control formulations". In: *Applied Mathematics and Computation.* 91(2): 259–284, 1998.

[65] Lasaint P and Raviart P. "On a Finite Element Method for Solving the Neutron Transport Equation". In: *Mathematical Aspects of Finite Elements in Partial Differential Equations.* Ed. by de Boor C. Academic Press, 1974, pp. 89–123.

[66] Leimkuhler B and Reich S. *Simulating Hamiltonian Dynamics.* Cambridge University Press, 2004.

[67] Lèvine J. *Analysis and Control of Nonlinear Systems - A Flatness-based Approach.* Springer, 2009.

[68] Leyendecker S. "Mechanical integrators for constrained dynamical systems in flexible multibody dynamics". 2006.

[69] Magnus K. *Kreisel: Theorie und Anwendungen.* 1971.

[70] Marsden J and Ratiu T. *Introduction to Mechanics and Symmetry.* 2nd. Springer, 1999.

[71] Marsden J and West M. "Discrete mechanics and variational integrators". In: *Acta Numerica.* 357–514, 2001.

[72] Massau J. *Mèmoire sur l'integration graphique des équations aux dérivées partielles.* F. Meyer-Van Loo, Gand, 1899.

[73] Mattsson S and Söderlind G. "Index Reduction in Differential-Algebraic Equations Using Dummy Derivatives". In: *SIAM J. Sci. Comput.* 14: 677–692, 1993.

[74] Murray R. "Trajectory Generation For a Towed Cable System Using Differential Flatness". In: IFAC World Congress. San Francisco, July 1996, pp. 395–400.

[75] Otto S and Seifried R. "Real-time trajectory control of an overhead crane using servo-constraints". In: *Multibody System Dynamics.* 42(1): 1–17, 2018.

[76] Petit N and Rouchon P. "Flatness of heavy chain systems". In: *SIAM Journal on Control and Optimization.* 40(2): 475–495, 2001.

[77] Reissner E. "On one-dimensional finite-strain beam theory: The plane problem". In: *Journal of Applied Mathematics and Physics (ZAMP).* 23: 795–804, 1972.

[78] Reissner E. "On finite deformations of space-curved beams". In: *Journal of Applied Mathematics and Physics (ZAMP).* 32: 734–744, 1981.

[79] Rosen A. "Applying the Lagrange Method to Solve Problems of Control Constraints". In: *Journal of Applied Mechanics.* 66(4): 1013–1015, Dec. 1999.

[80] Saadé C, Lejeunes S, Eyheramendy D, and Saad R. "Space-Time Isogeometric Analysis for linear and non-linear elastodynamics". In: *Computers and Structures.* 254: 106594, 2021.

[81] Sanz-Serna J. "On finite elements simultaneously in space and time". In: *International Journal for Numerical Methods in Engineering.* 19(4): 623–624, 1983.

[82] Sastry S. *Nonlinear systems: analysis, stability and control. Interdisciplinary applied mathematics.* Springer, New York, 1999.

[83] Sauer R. "Über die Gleitkurvennetze der ebenen plastischen Spannungsverteilungen bei beliebigem Fliebgesetz". In: *Zeitschrift für angewandte Mathematik und Mechanik.* 29(9): 274–279, 1949.

[84] Sauer R. *Anfangswertprobleme bei partiellen Differentialgleichungen.* Springer-Verlag, 1958.

[85] Seifried R. “Integrated mechanical and control design of underactuated multibody systems”. In: *Nonlinear Dynamics.* 67: 1539–1557, 2012.

[86] Seifried R. *Dynamics of Underactuated Multibody Systems.* Vol. 205. Solid Mechanics and its Applications. Springer-Verlag, 2014.

[87] Seifried R and Blajer W. “Analysis of servo-constraint problems for underactuated multibody systems”. In: *Mechanical Sciences.* 4(1): 113–129, 2013.

[88] Simo J. “A finite strain beam formulation. The three-dimensional dynamic problem. Part I”. In: *Computer Methods in Applied Mechanics and Engineering.* 49(1): 55–70, 1985.

[89] Simo J and Vu-Quoc L. “A three-dimensional finite-strain rod model. Part II: Computational aspects”. In: *Computer Methods in Applied Mechanics and Engineering.* 58(1): 79–116, 1986.

[90] Sira-Ramírez H and Agrawal S. *Differentially Flat Systems.* CRC Press, 2004.

[91] Stander N and Stein E. “An energy-conserving planar finite beam element for dynamics of flexible mechanisms”. In: *Engineering Computations.* 13(6): 60–85, 1996.

[92] Strang G. *Differential Equations and Linear Algebra.* Wellesley Cambridge Press, 2015.

[93] Ströhle T and Betsch P. “A simultaneous space-time discretization approach to the inverse dynamics of geometrically exact strings”. In: *International Journal for Numerical Methods in Engineering.* 123(11): 2573–2609, 2022.

[94] Ströhle T and Betsch P. *Galerkin space-time discretization approach to the inverse dynamics of flexible multibody systems.* Centre International de Rencontres Mathématiques (CIRM): Conference on Energy-Based Modeling, Simulation, and Control of Complex Constrained Multiphysical Systems. Luminy, Marseille, France, April 18-22, 2022.

[95] Timoshenko S. “LXVI. On the correction for shear of the differential equation for transverse vibrations of prismatic bars”. In: *The London, Edinburgh, and Dublin Philosophical Magazine and Journal of Science.* 41(245): 744–746, 1921.

[96] Timoshenko S. “X. On the transverse vibrations of bars of uniform cross-section”. In: *The London, Edinburgh, and Dublin Philosophical Magazine and Journal of Science.* 43(253): 125–131, 1922.

[97] v.Mises R. “Three remarks on the theory of the ideal plastic body”. In: *Reissner Anniversary Volume: Contributions to Applied Mechanics.* 415–429, 1949.

[98] Yang Y, Betsch P, and Zhang W. “Numerical integration for the inverse dynamics of a large class of cranes”. In: *Multibody System Dynamics.* 48(1): 1–40, 2020.

[99] Zienkiewicz O, Taylor R, and Zhu J. *The finite element method: Its basis and fundamentals.* Sixth Edition. Butterworth-Heinemann, 2005.

Schriftenreihe des Instituts für Mechanik
ISSN 2363-4936

Hrsg. von: Prof. Dr.-Ing. habil. Peter Betsch, Prof. Dr.-Ing. habil. Thomas Seelig

---

Band 1 Marlon Franke
Discretisation techniques for large deformation computational contact elastodynamics.
ISBN 978-3-7315-0278-4

Band 2 Philipp Hempel
Constitutive modeling of amorphous thermoplastic polymers with special emphasis on manufacturing processes.
ISBN 978-3-7315-0550-1

Band 3 Martin Helbig
Mehrskalenmodellierung von Schädigung in gummimodifizierten thermoplastischen Kunststoffen.
ISBN 978-3-7315-0571-6

Band 4 Maik Dittmann
Isogeometric analysis and hierarchical refinement for multi-field contact problems.
ISBN 978-3-7315-0616-4

Band 5 Yinping Yang
Numerical methods for the inverse dynamics simulation of underactuated mechanical systems.
ISBN 978-3-7315-0626-3

Band 6 Alexander Janz
Structure-preserving space-time discretization in a mixed framework for multi-field problems in large strain elasticity.
ISBN 978-3-7315-0926-4

Band 7 Mark Georg Schiebl
Thermodynamically consistent space-time discretization of non-isothermal mechanical systems in the framework of GENERIC.
ISBN 978-3-7315-1117-5

Band 8 Jonas Hund
Characterisation and modelling of PC/ABS blends.
ISBN 978-3-7315-1157-1

Die Bände sind unter www.ksp.kit.edu als PDF frei verfügbar oder als Druckausgabe bestellbar.

Band 9 Jonathan Schulte
Multi-field modeling and simulation of fiber-reinforced polymers.
ISBN 978-3-7315-1251-6

Band 10 Robin Pfefferkorn
EAS Elements for Solid Mechanics – Mesh Distortion Insensitive and Hourglassing-Free Formulations with Increased Robustness.
ISBN 978-3-7315-1271-4

Band 11 Timo Ströhle
Inverse dynamics of underactuated flexible mechanical systems governed by quasi-linear hyperbolic partial differential equations.
ISBN 978-3-7315-1336-0

Die Bände sind unter www.ksp.kit.edu als PDF frei verfügbar oder als Druckausgabe bestellbar.